CRISC® Review Manual

7th Edition

About ISACA

For more than 50 years, ISACA® (www.isaca.org) has advanced the best talent, expertise and learning in technology. ISACA equips individuals with knowledge, credentials, education and community to progress their careers and transform their organizations, and enables enterprises to train and build quality teams. Among those credentials, ISACA advances and validates business-critical skills and knowledge through the globally respected Certified Information Systems Auditor® (CISA®), Certified in Risk and Information Systems Control® (CRISC®), Certified Information Security Manager® (CISM®), Certified in the Governance of Enterprise IT® (CGEIT®) and Certified Data Privacy Solutions Engineer™ (CDPSE™) credentials. ISACA is a global professional association and learning organization that leverages the expertise of its 145,000 members who work in information security, governance, assurance, risk and privacy to drive innovation through technology. It has a presence in 188 countries, including more than 220 chapters worldwide.

Disclaimer

ISACA has designed and created *CRISC® Review Manual* primarily as an educational resource to assist individuals preparing to take the CRISC certification exam. It was produced independently from the CRISC exam and the CRISC Certification Working Group, which has had no responsibility for its content. Copies of past exams are not released to the public and were not made available to ISACA for preparation of this publication. ISACA makes no representations or warranties whatsoever with regard to these or other ISACA publications assuring candidates' passage of the CRISC exam.

Reservation of Rights

© 2021 ISACA. All rights reserved. No part of this publication may be used, copied, reproduced, modified, distributed, displayed, stored in a retrieval system or transmitted in any form by any means (electronic, mechanical, photocopying, recording or otherwise) without the prior written authorization of ISACA.

ISACA

1700 E. Golf Road, Suite 400
Schaumburg, IL 60173, USA
Phone: +1.847.660.5505
Fax: +1.847.253.1755
Contact us: https://support.isaca.org
Website: www.isaca.org

Participate in the ISACA Online Forums: https://engage.isaca.org/onlineforums

Twitter: http://twitter.com/ISACANews
LinkedIn: www.linkedin.com/company/isaca
Facebook: www.facebook.com/ISACAGlobal
Instagram: www.instagram.com/isacanews/

ISBN 978-1-60420-850-4
CRISC® Review Manual 7th Edition
Printed in the United States of America

CRISC® Review Manual 7th Edition

ISACA is pleased to offer the seventh edition of the *CRISC® Review Manual*. The purpose of the manual is to provide CRISC candidates with information and references to assist in the preparation and study for the Certified in Risk and Information Systems Control® (CRISC) exam.

The *CRISC® Review Manual* is updated regularly to keep pace with rapid changes in the identification, assessment, response, monitoring and reporting of risk and information systems (IS) controls. As with previous manuals, the seventh edition is the result of contributions from many qualified authorities who have generously volunteered their time and expertise. We respect and appreciate their contributions and hope their efforts provide extensive educational value to CRISC manual readers.

The sample questions contained in this manual are designed to depict the type of questions typically found on the CRISC exam and to provide further clarity to the content presented in this manual. The CRISC exam is a practice-based exam. Simply reading the reference material in this manual will not properly prepare candidates for the exam. The self-assessment questions are included for guidance only. Scoring results do not indicate future individual exam success.

Certification has resulted in a positive impact on many careers, including worldwide recognition for professional experience and enhanced knowledge and skills. The Certified in Risk and Information Systems Control certification is designed for IT and business professionals who have hands-on experience with risk identification, risk assessment, risk response and risk and IS control monitoring and reporting. ISACA wishes you success with the CRISC exam.

Acknowledgments

The *CRISC® Review Manual 7th Edition* is the result of collective efforts of many volunteers. ISACA members from throughout the globe participated, generously offering their talent and expertise. This international team exhibited a spirit and selflessness that has become the hallmark of contributors to ISACA manuals. Their participation and insight are truly appreciated.

ISACA has begun planning the eighth edition of the *CRISC® Review Manual*. Volunteer participation drives the success of the manual. If you are interested in becoming a member of the select group of professionals involved in this global project, please visit engage.isaca.org to be notified of all upcoming CRISC-related opportunities.

Authors
Edward C. McCabe, CISM, CGEIT, CRISC, ISO 27001, ISMS LI, The Rubicon Advisory Group, USA
James C. Samans, CISA, CISM, CRISC, CDPSE, CBCP, CISSP, CIPT, CPP, PMP, The American Instititues for Research, USA

Expert Reviewers
Christopher Anthony, CISA, CRISC, CDPSE, Singapore
Anagha Apte, CISA, CISM, CGEIT, CRISC, USA
Ercüment Ari, CISA, CISM, CRISC, CDPSE, CIPM, CIPP/E, FIP, ISO 22301/27001/27701 LA, Turkey
David Astles, CISM, CRISC, UK
Suvodeep Banerjee, CISM, CRISC, Philips, India
Ajay C. Bhayani, CRISC, AmbiSure Technologies Pvt. Ltd., India
Anestis Dimopoulos, CISA, CRISC, CGEIT, CIA, CIPM, CISSP, Baker Tilly, Greece
Adham Etoom, CISM, CRISC, FAIR, GCIH, PMP, Government of Jordan, Jordan
Urs Fischer, CISA, CRISC, CPA (Swiss), UBS Business Solutions AG, Switzerland
Sandra Fonseca, CISA, CISM, CRISC, CDPSE, USA
Andrew Foo, CISA, CISM, CGEIT, CRISC, CDPSE, CBCP, CCSK, CEH, CIPM, CISSP, CSM, CTIA, ECSA, ITIL MP, PMP, China
Jack Freund, Ph.D., CISA, CISM, CGEIT, CRISC, CDPSE, CSX-P, VisibleRisk, USA
Miriam Feito Huertas, CISA, CRISC, Spain
Yalcin Gerek, CISA, CGEIT, CRISC, CDPSE, TAC AS, Turkey
Peter Gwee, CISA, CISM, CRISC, CDPSE, Singapore
Mike Hughes, CISA, CISM, CGEIT, CRISC, CDPSE, MIoD, Prism RA, UK
Hideko Igarashi, CISA, CISM, CRISC, CDPSE, CISSP, Japan
Björg Ýr Jóhannsdóttir, CISA, CRISC, Borgun, Iceland
Leighton R. Johnson III, CISA, CISM, CRISC, CGEIT, CDPSE, ISO 27005 LRM, ISFMT, Inc., USA
Shamik Kacker, CISM, CRISC, CCSK, CCSP, CISSP, ITIL Expert, TOGAF 9 Foundation, Dell Corp., USA
Ramaswami Karunanithi, MBA, CISA, CGEIT, CRISC, CAMS, CBCI, CCSK, CCSP, CFE, CFSA, CGAP, CGMA (USA), CHFI, CIA, CIPP/E, CISSP, CMA (India & USA), CPA (Australia & USA), CPRM, CSCA, CSFX, CRMA, FCS, GRCA, GRCP, Lead Auditor ISO 27001, PBA, PMP, Prince2 Practitioner, RMP, NSW Government, Australia
Shruti Shrikant Kulkarni, CISA, CRISC, CCSK, CISSP, ITIL v3 Expert, Interpublic Group, UK
Emiko Kurihara, CRISC, CDPSE, CSA, CAIS-Lead Auditor, NISC, Japan
Christina Kwong, CRISC, USA
Salah Eddine Mahrach, CISA, CRISC, CDPSE, Moroccan Agency for Sustainable Energy, Morocco

Acknowledgments (cont.)

Adel Abdel Moneim, MBA, CISA, CISM, CGEIT, CRISC, CDPSE, CCFP-EU, CCISO, CCSK, CCSP, CDPO, CEH, CFR, CHFI, CISSP, CLSSP, CND, CSA, CTIA, ECES, ECIH, ECSA, EDRP, IoTSP, ISO 22301 SLA/SLI, ISO24762 LDRM, ISO27005 LRM, ISO27032 LCM, ISO27035 LIM, ISO27701 SLI/SLA, ISO29100 SLPI, ISO38500 LITCGM, LPT, Master ISO27001, MCCT, PECB MS Auditor, SABSA-SCF, TOGAF, ITU-ARCC, Egypt
Nnamdi Nwosu, CISA, CISM, CGEIT, CRISC, CEH, Fidson Healthcare Plc., Nigeria
Dapo Ogunkola, CISA, CRISC, CFE CFSA ACA, Ernst & Young London, UK
Ganiyu B. Oladimeji, CISA, CISM, CRISC. Moshood Abiola Polytechnic, Nigeria
Mercy Omollo, CISA, CRISC, CDPSE, Liberty Holdings, Kenya
Justin Orcutt, CRISC, CDPSE, USA
Teju Oyewole, CISA, CISM, CRISC, CDPSE, C|CISO, CCSP, CISSP, PMP, Canada
Vaibhav Patkar, CISA, CISM, CGEIT, CRISC, CDPSE, India
Anand Ramachandran, CISA, CISM, CRISC, SABSA Foundation SCF, Capgemini, UK
Andreas Schober, CISA, CISM, CGEIT, CRISC, CIA, CRMA, A1 Telekom Austria AG, Austria
Srinivasan Shamarao, CISA, CGEIT, CRISC, CDPSE, ACA, CIA, SSP Academy, India
Johan Sleeckx, CISA, CISM, CRISC, CDPSE, CISSP, SABASA SCF, Belgium
Ashish Vashishtha, CISA, CISM, CRISC, CDPSE, USA
Manoj Wadhwa, CISA, CISM, CRISC, EY, Singapore
Prometheus Yang, CISA, CISM, CRISC, Standard Chartered (Hong Kong) Bank, Hong Kong
Marcus Yin, CISA, CISM, CGEIT, CRISC, CISSP, CSM, PMP, PMI-RMP, RIMS-CRMP, Singapore
Guodong Zou, CISA, CISM, CGEIT, CRISC, CBAP, CBRM, Change Management Practitioner, PfMP, PgMP, PMP, Six Sigma Black Belt, Cisco Systems, China

New—CRISC Job Practice

Beginning in 2021, the Certified in Risk and Information Systems Control (CRISC) exam will test the new CRISC job practice.

An international job practice analysis is conducted periodically to maintain the validity of the CRISC certification program. A new job practice forms the basis of the CRISC beginning in 2021.

The primary focus of the job practice is the current tasks performed and the knowledge used by CRISCs. By gathering evidence of the current work practice of CRISCs, ISACA ensures that the CRISC program continues to meet the high standards for the certification of professionals throughout the world.

The findings of the CRISC job practice analysis are carefully considered and directly influence the development of new test specifications to ensure that the CRISC exam reflects the most current best practices.

The new job practice reflects the areas of study to be tested. The complete CRISC job practice can be found at https://www.isaca.org/credentialing/crisc.

Old CRISC Job Practice	New CRISC Job Practice
IT Risk Identification (27%)	Governance (26%)
IT Risk Assessement (28%)	IT Risk Assessment (20%)
Risk Response and Mitigation (23%)	Risk Response and Reporting (32%)
Risk and Control Monitoring and Reporting (22%)	Information Technology and Security (22%)

Table of Contents

About This Manual 15
Overview 15
Organization of This Manual 15
Format of This Manual 15
Preparing for the CRISC Exam 16
    Getting Started 16
    Using the CRISC Review Manual 16
        Features in the Review Manual 16
    Types of Questions on the CRISC Exam 17
    Using the CRISC Review Manual and Other ISACA Resources 17
    About the CRISC Review Questions, Answers & Explanations Manual 17
    About the CRISC Review Questions, Answers & Explanations Database 18

Chapter 1:
Governance 19

Overview 20
Domain 1 Exam Content Outline 20
Learning Objectives/Task Statements 20
Suggested Resources for Further Reading 21

Part A: Organizational Governance 26
    1.1 Organizational Strategy, Goals and Objectives 27
        1.1.1   The Context of IT Risk Management 29
        1.1.2   Key Concepts of Risk 31
        Example of Risk 32
        1.1.3   Importance and Value of IT Risk Management 32
        1.1.4   The IT Risk Strategy of the Business 33
        Types of IT-related Business Risk 33
        1.1.5   Alignment With Business Goals and Objectives 34
    1.2 Organizational Structure, Roles and Responsibilities 35
        1.2.1   RACI (Responsible, Accountable, Consulted, Informed) 35
        1.2.2   Key Roles 37
        1.2.3   Organizational Structure and Culture 37
    1.3 Organizational Culture 38
        1.3.1   Organizational Culture and Behavior and the Impact on Risk Management 39
        Culture 40
        Risk Awareness 40
        1.3.2   Risk Culture 41
        1.3.3   A Risk-driven Business Approach 43
        1.3.4   The Value of Risk Communication 44
    1.4 Policies and Standards 46
        1.4.1   Policies 46
        1.4.2   Standards 47
        1.4.3   Procedures 47
        1.4.4   Exception Management 49
        1.4.5   Risk Management Standards and Frameworks 49
        ISO 31000:2018 Risk Management — Guidelines 49
        COBIT and Information Risk 50
        IEC 31010:2019 Risk Management—Risk Assessment Techniques 50
        ISO/IEC 27001:2013 Information Technology—Security Techniques—Information Security Management Systems—Requirements 51
        ISO/IEC 27005:2018 Information Technology—Security Techniques—Information Security Risk Management 51
        NIST Special Publications 51

TABLE OF CONTENTS

   Example of a Risk Management Program Based on ISO/IEC 27005 52
 1.5 Business Process Review 53
  1.5.1 Risk Management Principles, Processes and Controls 55
   Principles of Risk Management 55
   Processes and Controls 56
  1.5.2 IT Risk in Relation to Other Business Functions 57
   Risk and Business Continuity 57
   Risk and Audit 57
   Risk and Information Security 58
   Control Risk 58
   Project Risk 59
   Change Risk 59
 1.6 Organizational Assets 59
  1.6.1 People 60
  1.6.2 Technology 60
  1.6.3 Data 60
  1.6.4 Intellectual Property 61
  1.6.5 Asset Valuation 62
   Asset Inventory and Documentation 62

Part B: Risk Governance 63
 1.7 Enterprise Risk Management and Risk Management Frameworks 63
  1.7.1 IT Risk Management Good Practices 63
  1.7.2 Establishing an Enterprise Approach to Risk Management 64
   Executive Sponsorship (Tone at the Top) 64
   Policy 64
 1.8 Three Lines of Defense 64
  1.8.1 The First Line of Defense: Operational Management 65
  1.8.2 The Second Line of Defense: Risk and Compliance Functions 65
  1.8.3 The Third Line of Defense: Audit 66
  1.8.4 The Role of the Risk Practitioner within the Three Lines of Defense 66
 1.9 Risk Profile 67
 1.10 Risk Appetite, Tolerance and Capacity 68
 1.11 Legal, Regulatory and Contractual Requirements 70
 1.12 Professional Ethics of Risk Management 71

Chapter 2:
IT Risk Assessment 73

Overview 74
Domain 2 Exam Content Outline 74
Learning Objectives/Task Statements 74
Suggested Resources for Further Study 75

Part A: IT Risk Identification 79
 2.1 Risk Events 80
  2.1.1 Risk Factors 81
  2.1.2 Methods of Risk Identification 84
  2.1.3 Changes in the Risk Environment 85
   Operational Integrity 85
   Industry Trends 86
   Forecasting Risk 86
 2.2 Threat Modeling and Threat Landscape 86
  2.2.1 Internal Threats 87
  2.2.2 External Threats 88
  2.2.3 Emerging Threats 89
  2.2.4 Additional Sources for Threat Information 89

TABLE OF CONTENTS

		Conducting Interviews	89
		Media Reports	90
		Observation	90
		Logs	91
		Self-assessments	91
		Third-party Assurance	91
		User Feedback	91
		Vendor Reports	91
	2.2.5	Threat, Misuse and Abuse-Case Modeling	92
		Threat Modeling	93
2.3	**Vulnerability and Control Deficiency Analysis**		**95**
	2.3.1	Sources of Vulnerabilities	96
		Network Vulnerabilities	96
		Physical Access	96
		Applications and Web-facing Services	97
		Utilities	97
		Supply Chain	98
		Equipment	98
		Cloud Computing	98
		Big Data	101
	2.3.2	Gap Analysis	102
	2.3.3	Vulnerability Assessment and Penetration Testing	103
		False Positives and Zero-Day Exploits	104
	2.3.4	Root Cause Analysis	106
2.4	**Risk Scenario Development**		**106**
	2.4.1	Risk Scenario Development Tools and Techniques	107
		Top-down Approach	109
		Bottom-up Approach	109
	2.4.2	Benefits of Using Risk Scenarios	110
	2.4.3	Developing IT Risk Scenarios	110
	2.4.4	Analyzing Risk Scenarios	113

Part B: IT Risk Analysis, Evaluation and Assessment 117

2.5	**Risk Assessment Concepts, Standards and Frameworks**		**117**
	2.5.1	Risk Ranking	119
		Risk Maps	120
	2.5.2	Risk Ownership and Accountability	121
	2.5.3	Documenting Risk Assessments	122
	2.5.4	Addressing Risk Exclusions	122
2.6	**Risk Register**		**123**
2.7	**Risk Analysis Methodologies**		**125**
	2.7.1	Quantitative Risk Assessment	125
	2.7.2	Qualitative Risk Assessment	126
	2.7.3	Semiquantitative/Hybrid Risk Assessment	126
2.8	**Business Impact Analysis**		**127**
	2.8.1	Business Continuity and Organizational Resiliency	128
	2.8.2	Regulatory and Contractual Obligations	129
	2.8.3	Strategic Investments	129
	2.8.4	Beyond Business Impact	129
	2.8.5	Business Impact Analysis and Risk Assessment	131
2.9	**Inherent, Residual and Current Risk**		**131**
	2.9.1	Inherent Risk	132
	2.9.2	Residual Risk	132
	2.9.3	Current Risk	132

CRISC® Review Manual 7th Edition
ISACA. All Rights Reserved.

TABLE OF CONTENTS

Chapter 3:
Risk Response and Reporting 133

Overview 134
Domain 3 Exam Content Outline 134
Learning Objectives/Task Statements 134
Suggested Resources for Further Study 135

Part A: Risk Response 138
  3.1 **Risk and Control Ownership** 138
    3.1.1 Ownership and Accountability 139
  3.2 **Risk Treatment/Risk Response Options** 140
    3.2.1 Aligning Risk Response with Business Objectives 140
    3.2.2 Risk Response Options 140
       Risk Acceptance 140
       Risk Mitigation 141
       Risk Transfer/Sharing 142
       Risk Avoidance 143
    3.2.3 Choosing a Risk Response 143
  3.3 **Third-party Risk Management** 144
  3.4 **Issue, Finding and Exception Management** 146
    3.4.1 Configuration Management 146
    3.4.2 Release Management 146
    3.4.3 Exception Management 147
    3.4.4 Change Management 147
    3.4.5 Issue and Finding Management 147
  3.5 **Management of Emerging Risk** 148
    3.5.1 Vulnerabilities Associated with New Controls 148
    3.5.2 Impact of Emerging Technologies on Design and Implementation of Controls 148

Part B: Control Design and Implementation 151
  3.6 **Control Types, Standards and Frameworks** 151
    3.6.1 Control Standards and Frameworks 153
    3.6.2 Administrative, Technical and Physical Controls 153
    3.6.3 Capability Maturity Models 154
  3.7 **Control Design, Selection and Analysis** 155
    3.7.1 Control Design and Selection 155
  3.8 **Control Implementation** 156
    3.8.1 Changeover (Go-live) Techniques 157
       Parallel Changeover 157
       Phased Changeover 158
       Abrupt Changeover 159
       Challenges Related to Data Migration 160
       Fallback (Rollback) 160
    3.8.2 Post-implementation Review 160
    3.8.3 Control Documentation 161
  3.9 **Control Testing and Effectiveness Evaluation** 161
    3.9.1 Good Practices for Testing 162
       Data 162
       Unit Testing and Code Review 163
       Quality Assurance 164
       Testing for Non-technical Controls 164
    3.9.2 Updating the Risk Register 165

Part C: Risk Monitoring and Reporting 167
  3.10 **Risk Treatment Plans** 167
  3.11 **Data Collection, Aggregation, Analysis and Validation** 170

TABLE OF CONTENTS

  3.11.1 Data Collection and Extraction Tools and Techniques 171
    Logs 172
    Security Information and Event Management 173
    Integrated Test Facilities 174
    External Sources of Information 175
3.12 Risk and Control Monitoring Techniques 175
  3.12.1 Monitoring Controls 175
  3.12.2 Control Assessment Types 177
    Self-assessment 177
    IS Audit 177
    Vulnerability Assessment 177
    Penetration Testing 178
    Third-party Assurance 179
3.13 Risk and Control Reporting Techniques 179
  3.13.1 Heat maps 179
  3.13.2 Scorecards 180
  3.13.3 Dashboards 181
3.14 Key Performance Indicators 182
3.15 Key Risk Indicators 183
  3.15.1 KRI Selection 184
  3.15.2 KRI Effectiveness 184
  3.15.3 KRI Optimization 185
  3.15.4 KRI Maintenance 185
  3.15.5 Using KPIs with KRIs 185
3.16 Key Control Indicators 186

Chapter 4:
Information Technology and Security 189

Overview 190
Domain 4 Exam Content Outline 190
Learning Objectives/Task Statements 190
Suggested Resources for Further Study 191

Part A: Information Technology Principles 195
 4.1 Enterprise Architecture 195
  4.1.1 Maturity Models 196
 4.2 IT Operations Management 197
  4.2.1 Hardware 197
    Supply Chain Management 198
  4.2.2 Software 198
    Operating Systems 199
    Applications 200
    Databases 200
    Software Utilities 201
  4.2.3 Environmental Controls 201
  4.2.4 Networks 202
    Protocols 203
    Cabling 205
    Repeaters 206
    Switches 206
    Routers 206
    Firewalls 207
    Proxies 208
    Intrusion Systems 208
    The Domain Name System 208
    Wireless Access Points 209
    Network Architecture 209

TABLE OF CONTENTS

  Network Topologies 210
  Network Types 211
  Software-defined Networking 212
  Demilitarized Zones 212
  Virtual Private Networks 213
 4.2.5 Technology Refresh 213
 4.2.6 IT Operations and Management Evaluation 214
  Configuration Management 214
 4.2.7 Virtualization and Cloud Computing 214
4.3 Project Management 216
 4.3.1 Project Risk 216
 4.3.2 Project Closeout 218
4.4 Enterprise Resiliency 218
 4.4.1 Business Continuity 219
  Recovery Objectives 220
 4.4.2 Disaster Recovery 220
4.5 Data Life Cycle Management 221
 4.5.1 Data Management 222
 4.5.2 Data Loss Prevention 224
4.6 System Development Life Cycle 224
4.7 Emerging Trends in Technology 226
 4.7.1 Omnipresent Connectivity 227
  Bring Your Own Devices (BYOD) 227
  The Internet of Things 228
 4.7.2 Massive Computing Power 229
  Decryption 229
  Deepfakes 229
  Big Data 230
 4.7.3 Blockchain 230
 4.7.4 Artificial Intelligence 230

Part B: Information Security Principles 233
4.8 Information Security Concepts, Frameworks and Standards 234
 4.8.1 Likelihood and Impact 234
 4.8.2 CIA Triad 234
  Confidentiality 235
  Integrity 236
  Availability 236
  Nonrepudiation 236
  System Authorization 236
 4.8.3 Segregation of Duties 237
 4.8.4 Cross-training and Job Rotation 237
 4.8.5 Access Control 238
  Identification 238
  Authentication 238
  Authorization 239
  Accountability 240
 4.8.6 Encryption 240
  Asymmetric Algorithms 242
  Message Integrity and Hashing Algorithms 242
  Digital Signatures 243
  Certificates 244
  Public Key Infrastructure 245
  Disadvantages of Encryption 245
  Summary of the Core Concepts of Cryptography 245
4.9 Information Security Awareness Training 246

TABLE OF CONTENTS

4.10 Data Privacy and Principles of Data Protection 247
    4.10.1 Key Concepts of Data Privacy 248
        Informed Consent 248
        Privacy Impact Assessment 248
        Minimization 248
        Destruction 248
    4.10.2 Risk Management in a Privacy Context 248

Appendix A: CRISC General Exam Information 251
Requirements for Certification 251
Successful Completion of the CRISC Exam 251
Experience in Risk 251
Description of the Exam 251
Registration for the CRISC Exam 251
CRISC Program Accreditation Renewed Under ISO/IEC 17024:2012 252
Scheduling the Exam 252
Sitting for the Exam 252
    Budgeting Your Time 253
    Grading the Exam 253

Appendix B: CRISC Job Practice 255

Glossary 259

Page intentionally left blank

About This Manual

Overview

The *CRISC® Review Manual 7th Edition* is a reference guide designed to assist candidates in preparing for the CRISC examination. **The manual is one source of preparation for the exam but should not be thought of as the only source nor viewed as a comprehensive collection of all the information and experiences that are required to pass the exam.** No single source offers such coverage and detail.

As candidates read through the manual and encounter topics that are new to them or topics in which they feel their knowledge and experience are limited, additional references should be sought. The examination will be composed of questions testing the candidate's technical and practical knowledge and ability to apply the knowledge (based on experience) in given situations.

Organization of This Manual

The *CRISC Review Manual 7th Edition* is divided into four chapters covering the CRISC domains tested on the exam in the percentages listed below:

Domain 1	Governance	26 percent
Domain 2	IT Risk Assessment	20 percent
Domain 3	Risk Response and Reporting	32 percent
Domain 4	Information Technology and Security	22 percent

Note: Each chapter defines the tasks that CRISC candidates are expected to know how to do and includes a series of learning objectives required to perform those tasks. These constitute the current practices for the IT risk practitioner. The detailed CRISC job practice can be viewed at www.isaca.org/credentialing/crisc/crisc-job-practice-areas.

This manual has been developed and organized to assist in the study of these areas. Exam candidates should evaluate their strengths, based on knowledge and experience, in each of these areas.

Format of This Manual

Each chapter of the *CRISC Review Manual* follows the same format:
- The Overview section provides a summary of the focus of the chapter along with:
  - The domain exam content outline
  - Related task statements
  - Suggested resources for further study
  - Self-assessment questions
- The Content section includes:
  - Content to support the different areas of the job practice
  - Definitions of terms commonly found on the exam

Please note that the manual has been written using standard American English.

Submit suggestions to enhance the review manual or suggested reference materials to studymaterials@isaca.org.

Preparing for the CRISC Exam

The CRISC exam evaluates a candidate's practical knowledge experiences and application of the job practice domains as described in this Review Manual. We recommend that the exam candidate look to multiple resources to prepare for the exam, including this Review Manual, along with external publications. This section covers some tips for studying for the exam.

Read to understand the areas that need more knowledge. Then, see reference sources to expand those areas and, also, gain experience in those areas.

Actual exam questions test the candidate's practical application of this knowledge. The sample self-assessment questions and answers, with explanations, at the end of each chapter introduce question structure and general content. Remember that the sample questions are not actual exam items, but are similar to items that may appear on the exam. Use reference material to find other publications to better understand detailed information on the topics addressed in this manual.

Getting Started

Having adequate time to prepare for the CRISC exam is critical. Most candidates spend between three and six months studying prior to taking the exam. Set aside a designated time each week to study and increase study time as the exam date approaches.

It helps to develop a plan for studying to prepare for the exam.

Using the CRISC Review Manual

The *CRISC Review Manual* is divided into four chapters, each corresponding to a domain in the CRISC job practice. While the *Review Manual* does not include every concept that could be tested on the CRISC exam, it does cover a breadth of knowledge to provide a solid base for the exam candidate. The manual is one source of preparation for the exam and should not be thought of as the only source, nor viewed as a comprehensive collection of all the information and experience that are required to pass the exam.

Features in the Review Manual

The CRISC Review Manual includes several features to help you navigate the job practice and enhance your retention of the material.

Review Manual Feature	Description
Overview	The overview provides the context of the domain, including the job practice areas and applicable learning objectives and task statements.
Suggested Resources for Further Study	Because many of the concepts presented within the review manual are complex, candidates may find it useful to refer to external sources to supplement their understanding of those concepts. The suggested references for each chapter are resources intended to help enhance study efforts.

Self-assessment Questions and Answers	The self-assessment questions in each chapter are not intended to measure the candidate's ability to answer questions correctly on the CRISC exam for that area.
	The questions are intended to familiarize the candidate with question structure and they may or may not be similar to questions that appear on the actual examination.
Glossary	The glossary included at the end of the manual contains terms that apply to:
	• The material included in the chapters
	• Related areas not specifically discussed in the manual
	Since the glossary is an extension of the manual text, it can point to areas the candidate may want to explore further using additional references.

Types of Questions on the CRISC Exam

CRISC exam questions are developed with the intent of measuring and testing practical knowledge and the application of IT risk-management principles. As previously mentioned, all questions are presented in a multiple-choice format and are designed for one best answer.

The candidate is cautioned to read each question carefully. Knowing the types of questions are asked and how to study to answer them will go a long way toward answering them correctly. There can be many potential solutions to the scenarios posed in the questions, depending on industry, geographical location, etc. It is advisable to consider the information provided in the question and to determine the best answer of the options provided.

Each CRISC question has a stem (question) and four options (answer choices). The candidate is asked to choose the correct or best answer from the options. The stem may be in the form of a question or an incomplete statement. In some instances, a scenario or description also may be included. These questions normally include a description of a situation and require the candidate to answer two or more questions based on the information provided.

A helpful approach to these questions includes the following:

- Read the entire stem and determine what the question is asking. Look for key words such as "BEST," "MOST," "FIRST," etc. and key terms that may indicate what domain or concept that is being tested.
- Read all the options and then read the stem again to see if you can eliminate any of the options based on your immediate understanding of the question.
- Re-read the remaining options and bring in any personal experience to determine which is the best answer to the question.

Using the CRISC Review Manual and Other ISACA Resources

The *CRISC Review Manual* can be used in conjunction with other CRISC exam preparation activities. The following products are based on the CRISC job practice and referenced job practice areas can be used to find related content within the *CRISC Review Manual*. These resources include:

- *CRISC® Review Questions, Answers & Explanations Manual 6th Edition*
- *CRISC® Review Questions, Answers & Explanations Database—12 Month Subscription*
- CRISC Online Review Course
- CRISC review courses (provided by local ISACA chapters and accredited training organizations)

About the CRISC Review Questions, Answers & Explanations Manual

Candidates may also wish to enhance their study and preparation for the exam by using the *CRISC® Review Questions, Answers & Explanations Manual 6th Edition*.

ABOUT THIS MANUAL

The *CRISC Review Questions, Answers & Explanations Manual 6th Edition* consists of 600 multiple-choice study questions, answers and explanations arranged in the domains of the current CRISC job practice.

Questions in this publication are representative of the types of questions that could appear on the exam and include explanations of the correct and incorrect answers. Questions are sorted by the CRISC domains and as a sample test. These publications are ideal for use in conjunction with the *CRISC Review Manual 7th Edition*. These manuals can be used as study sources throughout the study process or as part of a final review to determine where candidates may need additional study. It should be noted that these questions and suggested answers are provided as examples; they are not actual questions from the examination and may differ in content from those that actually appear on the exam.

About the CRISC Review Questions, Answers & Explanations Database

Another study aid that is available is the CRISC® Review Questions, Answers & Explanations Database – 12 Month Subscription. The online database consists of the 600 questions, answers and explanations included in the *CRISC Review Questions, Answers & Explanations Manual 6th Edition*. With this product, CRISC candidates can quickly identify their strengths and weaknesses by taking random sample exams of varying length and breaking the results down by domain. Sample exams also can be chosen by domain, allowing for concentrated study, one domain at a time, and other sorting features such as the omission of previous correctly answered questions are available.

Note: When using the CRISC review materials to prepare for the exam, it is important to note that they cover a broad spectrum of IT-related business risk and IS control issues. Candidates should not assume that reading these manuals and answering review questions will fully prepare them for the examination. Since actual exam questions often relate to practical experiences, candidates should refer to their own experiences and other reference sources and draw on the experiences of colleagues and others who have earned the CRISC designation.

Chapter 1:
Governance

Overview

Domain 1 Exam Content Outline 20
Learning Objectives/Task Statements 20
Suggested Resources for Further Reading 21

Part A: Organizational Governance

1.1 Organizational Strategy, Goals and Objectives 27
1.2 Organizational Structure, Roles and Responsibilities 35
1.3 Organizational Culture 38
1.4 Policies and Standards 46
1.5 Business Process Review 53
1.6 Organizational Assets 59

Part B: Risk Governance

1.7 Enterprise Risk Management and Risk Management Frameworks 63
1.8 Three Lines of Defense 64
1.9 Risk Profile 67
1.10 Risk Appetite, Tolerance and Capacity 68
1.11 Legal, Regulatory and Contractual Requirements 70
1.12 Professional Ethics of Risk Management 71

Overview

At its core, governance is the ability to meet stakeholder needs by providing value. This is achieved through the proper balancing of both performance and conformance requirements defined by the enterprise and only accomplished by ensuring that a proper risk-management capability is in place. Having a well-defined risk-management program ensures that enterprises are able to identify, analyze, evaluate, assess and respond to those threats that pose the greatest risk . This allows enterprises to prioritize their limited resources, realize benefits and ultimately deliver value to stakeholders. Effective risk management bridges the requirements for performance and conformance and establishes sound governance principles and practices.

This domain represents 26 percent (approximately 39 questions) of the exam.

Domain 1 Exam Content Outline

A. Organizational Governance

1. Organizational Strategy, Goals and Objectives
2. Organizational Structure, Roles and Responsibilities
3. Organizational Culture
4. Policies and Standards
5. Business Processes
6. Organizational Assets

B. Risk Governance

1. Enterprise Risk Management and Risk Management Frameworks
2. Three Lines of Defense
3. Risk Profile
4. Risk Appetite and Risk Tolerance
5. Legal, Regulatory and Contractual Requirements
6. Professional Ethics of Risk Management

Learning Objectives/Task Statements

Upon the completion of this chapter, the risk practitioner should be able to:

1. Collect and review existing information regarding the organization's business and IT environments.
2. Identify potential or realized impacts of IT risk to the organization's business objectives and operations.
3. Identify threats and vulnerabilities to the organization's people, processes and technology.
4. Evaluate threats, vulnerabilities and risk to identify IT risk scenarios.
5. Establish accountability by assigning and validating appropriate levels of risk and control ownership.
6. Establish and maintain the IT risk register and incorporate it into the enterprise-wide risk profile.
7. Facilitate the identification of risk appetite and risk tolerance by key stakeholders.
8. Promote a risk-aware culture by contributing to the development and implementation of security awareness training.
9. Conduct a risk assessment by analyzing IT risk scenarios and determining their likelihood and impact.

10. Review the results of risk analysis and control analysis to assess any gaps between current and desired states of the IT risk environment.
11. Facilitate the selection of recommended risk responses by key stakeholders.
12. Collaborate with risk owners on the development of risk treatment plans.
13. Collaborate with control owners on the selection, design, implementation and maintenance of controls.
14. Define and establish key risk indicators (KRIs).
15. Monitor and analyze key risk indicators (KRIs).
16. Collaborate with control owners on the identification of key performance indicators (KPIs) and key control indicators (KCIs).

Suggested Resources for Further Reading

Chapman, Robert J.; *Simple Tools and Techniques for Enterprise Risk Management, Second Edition*, John Wiley & Sons Inc., USA, 2012

Committee of Sponsoring Organizations of the Treadway Commission; *Enterprise Risk Management: Understanding and Communicating Risk Appetite,* USA, 2012

Committee of Sponsoring Organizations of the Treadway Commission; *Enterprise Risk Management for Cloud Computing*, USA, 2012

Committee of Sponsoring Organizations of the Treadway Commission; *Enterprise Risk Management: Integrating with Strategy and Performance—Executive Summary,* USA, 2017

Freund, Jack; Jack Jones; *Measuring and Managing Information Risk*, Butterworth-Heinemann, USA, 2014

Jordan, Ernest; Luke Silcock; *Beating IT Risks*, John Wiley & Sons Inc., USA, 2005

Hubbard, Douglas; *How to Measure Anything: Finding the Value of Intangibles in Business, 3rd edition*, Wiley, USA, 2014

Hubbard, Douglas; *How to Measure Anything in Cybersecurity Risk*, Wiley, USA 2016

Hubbard, Douglas; *The Failure of Risk Management: Why It's Broken and How to Fix It*, Wiley, USA, 2009

ISACA, *COBIT Focus Area: Information Risk,* USA, 2021

ISACA, *COBIT 2019: Introduction and Methodology*, USA, 2018

ISACA, *Cybersecurity Fundamentals Study Guide 3rd Edition*, USA, 2021

ISACA, *The Risk IT Framework 2nd Edition*, USA, 2020

ISACA, T*he Risk IT Practitioner Guide 2nd Edition*, USA, 2020

National Institute of Standards and Technology (NIST), *NIST Special Publication 800-39: Managing Information Security Risk*, USA, 2011

Taleb, Nassim Nicholas; *The Black Swan Second Edition: The Impact of the Highly Improbable*, Random House, USA, 2010

CHAPTER 1— GOVERNANCE

Westerman, George; Richard Hunter; *IT Risk: Turning Business Threats Into Competitive Advantage*, Harvard Business School Press, USA, 2007

SELF-ASSESSMENT QUESTIONS

CRISC self-assessment questions support the content in this manual and provide an understanding of the type and structure of questions that have typically appeared on the exam. Questions are written in a multiple-choice format and are designed for one best answer. Each question has a stem (question) and four options (answer choices). The stem may be written in the form of a question or an incomplete statement. In some instances, a scenario or a description problem may also be included. These questions normally include a description of a situation and require the candidate to answer two or more questions based on the information provided. Many times, a question will require the candidate to choose the **MOST** likely or **BEST** answer among the options provided.

In each case, the candidate must read the question carefully, eliminate known incorrect answers and then make the best choice possible. Knowing the format in which questions are asked, and how to study and gain knowledge of what will be tested, will help the candidate correctly answer the questions.

1. Shortly after performing the annual review and revision of corporate policies, a risk practitioner becomes aware that a new law may affect security requirements for the human resources system. The risk practitioner should:

    A. analyze in detail how the law may affect the enterprise.
    B. ensure that necessary adjustments are made during the next review cycle.
    C. initiate an *ad hoc* revision of the corporate policy.
    D. notify the system custodian to implement changes.

2. Which of the following choices provides the **BEST** view of risk management?

    A. An interdisciplinary team.
    B. A third-party risk assessment service provider.
    C. The enterprise's IT department.
    D. The enterprise's internal compliance department.

3. Which of the following choices is a **PRIMARY** consideration when developing an IT risk awareness program?

    A. Why technology risk is owned by IT.
    B. How technology risk can impact each attendee's area of business.
    C. How business process owners can transfer technology risk.
    D. Why technology risk is more difficult to manage compared to other risk.

4. It is **MOST** important that risk appetite is aligned with business objectives to ensure that:

    A. resources are directed toward areas of low risk tolerance.
    B. major risk is identified and eliminated.
    C. IT and business goals are aligned.
    D. the risk strategy is adequately communicated.

5. The risk to an information system that supports a critical business process is owned by:

   A. the IT director.
   B. senior management.
   C. the risk management department.
   D. the system users.

Answers on page 24

CHAPTER 1 — GOVERNANCE

Chapter 1 Answer Key

Self-Assessment Questions

1. **A. Assessing how the law may affect the enterprise is the best course of action. The analysis must also determine whether existing controls already address the new requirements.**
   B. Ensuring that necessary adjustments are made during the next review cycle is not the best answer, particularly when the law does affect the enterprise. While an annual review cycle may be sufficient in general, significant changes in the internal or external environment should trigger an *ad hoc* reassessment.
   C. Corporate policy should be developed in a systematic and deliberate manner. An *ad hoc* amendment to the corporate policy is not warranted and may create risk rather than reducing it.
   D. Notifying the system custodian to implement changes is inappropriate. Changes to the system should be implemented only after approval by the process owner.

2. **A. Having an interdisciplinary team contribute to risk management ensures that all areas are adequately considered and included in the risk assessment processes to support an enterprise view of risk.**
   B. Engaging a third party to perform a risk assessment may provide additional expertise to conduct the risk assessment; but without internal knowledge, it will be difficult to assess the adequacy of the risk assessment performed.
   C. A risk assessment performed by the enterprise's IT department is unlikely to reflect the view of the entire enterprise.
   D. The internal compliance department ensures the implementation of risk responses based on the requirement of management. It generally does not take an active part in implementing risk responses for items that do not have regulatory implications.

3. A. IT does not own technology risk. An appropriate topic of IT risk awareness training may be the fact that many types of IT risk are owned by the business. One example may be the risk of employees exploiting insufficient segregation of duties within an enterprise resource planning system.
   B. Stakeholders must understand how the IT-related risk impacts the overall business.
   C. Transferring risk is not of primary consideration in developing a risk awareness program. It is a part of the risk response process.
   D. Technology risk may or may not be more difficult to manage than other types of risk. Although this is important from an awareness point of view, it is not as primary as understanding the impact in the area of business.

4. **A. Risk appetite is the amount of risk that an enterprise is willing to take on in pursuit of value. Aligning it with business objectives allows an enterprise to evaluate and deploy valuable resources toward those objectives where the risk tolerance (for loss) is low.**
   B. There is no link between aligning risk appetite with business objectives and identification and elimination of major risk, and although risk can typically be reduced to an acceptable level using various risk response options, its elimination is rarely cost-effective even when it is possible.
   C. Alignment of risk appetite with business objectives does converge IT and business goals to a point, but alignment is not limited to these two areas. Other areas include organizational, strategic and financial objectives, among other objectives.
   D. Communication of the risk strategy does not depend on aligning risk appetite with business objectives.

CHAPTER 1— GOVERNANCE

5.   A. The IT director manages the IT systems on behalf of the business owners.

   B. Senior management is responsible for the acceptance and mitigation of all risk.

   C. The risk management department determines and reports on level of risk but does not own the risk.

   D. The system users are responsible for utilizing the system properly and following procedures, but they do not own the risk.

CHAPTER 1— GOVERNANCE

Part A: Organizational Governance

Governance establishes requirements for how to achieve the proper balance of performance and conformance within an enterprise in order to meet stakeholder needs and deliver value. It also establishes the accountability for protection of organizational assets. In a corporate structure, the directors of an enterprise(frequently organized as a board) are accountable for governance and entrust the senior management team with the responsibility to manage the day-to-day operations in alignment with the strategic mandates that the directors approve. Similar arrangements exist within cooperative and partnership-style enterprises, although the actual titles may differ. **Figure 1.1** provides an overview of the risk governance structure.

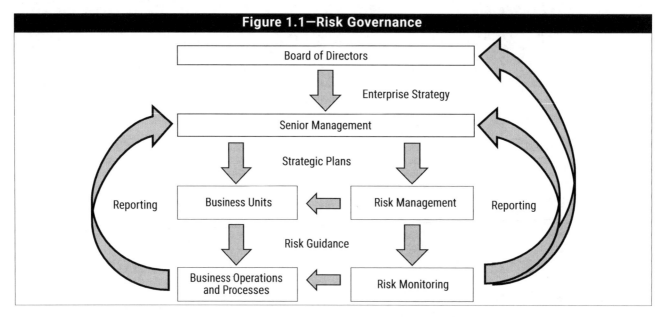

Governance is not confined to senior management and the board of directors; it is applicable to all departments within an enterprise. It may take the form of financial accountability and oversight, operational effectiveness, legal and regulatory compliance, adoption of fair labor practices, social responsibility and governance of IT investment, operations and control. Risk management is the keystone of governance. Accurate information is required to be able to correctly understand the various threats and subsequent risk being faced and how the enterprise chooses to respond.

The term "governance" has moved to the forefront of business thinking in response to examples of good governance on one end of the spectrum and global business failures resulting from poor governance on the other. Corporate governance is the system by which organizations evaluate, direct, monitor and ultimately control an enterprise.

By extension, the governance of enterprise IT is the system by which the current and future use of IT are evaluated, directed, monitored and ultimately controlled. The objective of any governance system is to enable enterprises to create value for their stakeholders or to promote the creation of value. Value creation, in turn, is comprised of benefit realization, risk optimization and resource optimization. As mentioned previously, risk optimization is essential for the success of any governance system and cannot be isolated, as an enterprise will not be able to properly plan and optimize resources, nor realize benefits, without proper governance and risk management.

Governance answers four questions:

- Are we doing the right things?
- Are we doing them the right way?
- Are we getting them done well?
- Are we seeing expected benefits?

CHAPTER 1— GOVERNANCE

There is a distinction between the governance and management functions. Management activities focus on planning, building, running and monitoring to ensure proper and continued alignment with the defined goals and objectives established by the governing body. These activities are performed to ensure the ability to achieve the enterprise's strategic vision, as directed by the governing body and resulting in creating value. A well-managed enterprise that lacks proper governance often develops and executes defined plans to meet objectives; however, because those objectives are not aligned with the enterprise's strategic vision and goals, they do not create any value.

As with enterprise governance, those who are responsible for governance over risk management must evaluate, direct and monitor risk management efforts. Senior management needs to identify, define and communicate the goals and objectives of risk management capabilities. This communication ensures that risk-management processes, practices and activities are properly aligned and become part of business processes and are embedded within the normal routine and operations within the enterprise.

Effective risk governance helps ensure that risk management practices are embedded in the enterprise, enabling it to secure optimal risk-adjusted return.

1.1 Organizational Strategy, Goals and Objectives

An enterprise exists for the sole purpose of achieving the defined strategic vision. An enterprise's strategy is the focus of its efforts; these are the primary drivers behind how investments and decisions are made and which actions are taken. So how does an enterprise identify its strategy? COBIT proposes the four strategic archetypes to help describe strategies (**figure 1.2**).

Figure 1.2—Enterprise Strategy Design Factor	
Strategy Archetype	Explanation
Growth/Acquisition	The enterprise has a focus on growing (revenues).[1]
Innovation/Differentiation	The enterprise has a focus on offering different and/or innovative products and services to their clients.[2]
Cost Leadership	The enterprise has a focus on short-term cost minimization.[3]
Client Service/Stability	The enterprise has a focus on providing stable and client-oriented service.[4]
Source: ISACA, *COBIT 2019: Introduction and Methodology*, USA, 2018, figure 4.5	

Understanding the strategy archetype allows the enterprise to align risk governance processes and ensure that risk efforts are evaluated, directed and monitored in the proper context. This requires the enterprise's governing body to ensure that the risk appetite and risk tolerance are understood and that they are communicated throughout the enterprise. This, in turn, establishes the goals of risk management capabilities. At a high-level, the goals for risk practitioners is to:

- Provide accurate, complete and timely information required to make the best, most well-informed decision possible by senior management.

---

[1] Corresponds with prospector in the Miles-Snow typology. See "Miles and Snow's Typology of Defender, Prospector, Analyzer, and Reactor," Elibrary, https://ebrary.net/3737/management/miles_snows_typology_defender_prospector_analyzer_reactor.
[2] See Reeves, Martin; Claire Love, Philipp Tillmanns, "Your Strategy Needs a Strategy," *Harvard Business Review*, September 2012, https://hbr.org/2012/09/your-strategy-needs-a-strategy, specifically regarding visionary and shaping.
[3] Corresponds to cost leadership; see University of Cambridge, "Porter's Generic Competitive Strategies (ways of competing)," Institute for Manufacturing (IfM) Management Technology Policy, https://www.ifm.eng.cam.ac.uk/research/dstools/porters-generic-competitive-strategies/. Also corresponds to operational excellence; see Treacy, Michael; Fred Wiersema, "Customer Intimacy and Other Value Disciplines," *Harvard Business Review*, January/February 1993, https://hbr.org/1993/01/customer-intimacy-and-other-value-disciplines
[4] Corresponds with defenders in the Miles-Snow typology. See *op cit* "Miles and Snow's Typology of Defender, Prospector, Analyzer, and Reactor."

CHAPTER 1— GOVERNANCE

- Identify, assess and advise, with the appropriate response, the risk that has the highest likelihood of occurrence and/or ability to impact the enterprise's ability to successfully achieve its goals and objectives.
- Allow for the ability to balance performance and conformance requirements that best suit the enterprise through the activities previously described.

To ensure the enterprise can deliver on its goals, clearly defined objectives are necessary. **Figure 1.3** describes the four core objectives of risk governance.

Figure 1.3—Risk Governance Objectives	
Objective	Description
1. Establish and maintain a common risk view.	Effective risk governance establishes the common view of risk for the enterprise. The risk governance function sets the tone of the business regarding how to determine an acceptable level of risk tolerance. Risk governance is a continuous life cycle that requires regular reporting and ongoing review. The risk governance function must oversee the operations of the risk management team.
2. Integrate risk management into the enterprise.	Integrating risk management into the enterprise enforces a holistic enterprise risk management (ERM) approach across the entire enterprise. It requires the integration of risk management into every department, function, system and geographic location. Understanding that risk in one department or system may pose an unacceptable risk to another department or system requires that all business processes be compliant with a baseline level of risk management. Therefore, it is important to provide both top-down and bottom-up approaches to enterprise risk. The objective of ERM is to establish the authority to require all business processes to undergo a risk analysis on a periodic basis or when there is a significant change to the internal or external environment.
3. Make risk-aware business decisions.	To make risk-aware business decisions, the risk governance function must consider the full range of opportunities and consequences of each decision and its impact on the enterprise, society and the environment.
4. Ensure that risk management controls are implemented and operating correctly.	Governance requires oversight and due diligence to ensure that the enterprise is following up on the implementation and monitoring of controls to ensure that the controls are effective to mitigate risk and protect organizational assets.

In order to deliver and provide a successful risk management capability, enterprises should:

- Consider the available methods, models and frameworks that best align with the enterprise's culture, maturity and industry
- Define the risk taxonomy (i.e., how risk is classified)
- Define the risk ontology (i.e., the set of concepts, categories and their properties and the relations of the various risk elements)
- Integrate risk efforts into the enterprise
- Manage risk within the enterprise
- Determine the process for making risk-based business decisions
- Track and trend the outcomes of risk efforts
- Report on status of risk being managed by the enterprise
- Allocate the necessary resources to implement IT risk management. (e.g., staff with the right experience and skills)

1.1.1 The Context of IT Risk Management

Risk management is defined as the coordinated activities, practices and processes that are used in an attempt to inform, direct and influence an enterprise with regard to risk.

In simple terms, risk can be viewed as a challenge to achieving objectives and risk management as the activity undertaken to predict challenges and lower their chances of occurring and their impact. Effective risk management can also assist in maximizing opportunities, and the risk practitioner should keep this upside/downside duality of risk in mind. For example, a risk decision might take the form of potential benefits that may accrue if opportunities are taken, versus missed benefits if those same opportunities are foregone.

The dual nature of risk is a result of its use in different contexts by business and IT, and it is not always easy to draw the distinction. *International Organization for Standardization (ISO) 31000:2018 – Risk Management Principles and Guidelines* calls risk "the effect of uncertainty on objectives. An effect is a deviation from the expected—positive and/or negative." However, *ISO/International Electrotechnical Commission (IEC) 27005:2018 – Information technology – Security techniques – Information security risk management* regards risk solely from a negative angle, stating "information security risk is the potential that a given threat will exploit vulnerabilities of an asset or group of assets and thereby cause harm to the organization." Risk practitioners who are able to effectively view risk from both perspectives are likely to find that they can more easily discuss risk with business and IT professionals without causing confusion.

Risk management starts with understanding the enterprise, but the risk practitioner should bear in mind that the enterprise is heavily influenced by the environment (internal and external) and/or context in which it operates. Assessing an enterprise's context includes evaluating the intent and capability of threats, the relative value of assets or resources and the trust that must be placed in them, and the presence and extent of vulnerabilities that might be exploited to intercept, interrupt, modify or fabricate data in information assets. Other factors to consider include:

- Dependency of the enterprise on third parties, such as a supply chain—especially when based in another geographic region or reliant upon just-in-time delivery
- Influence of financing, debt and partners or substantial stakeholders
- Vulnerability to changes in economic or political conditions
- Changes to market trends and patterns
- Emergence of new competition
- Emergence of new, disruptive technologies
- Impact of new legislation
- Existence of potential natural disaster
- Constraints caused by legacy systems and antiquated technology
- Strained labor relations and inflexible management

The strategy of the enterprise drives the individual lines of business that make up the enterprise, and each line of business will develop information systems that support its business function. **Figure 1.4** illustrates how IT risk relates to overall risk of the enterprise.

CHAPTER 1— GOVERNANCE

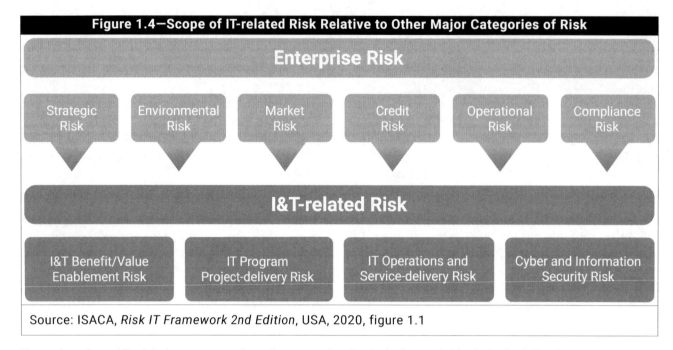

Figure 1.4—Scope of IT-related Risk Relative to Other Major Categories of Risk

Source: ISACA, *Risk IT Framework 2nd Edition*, USA, 2020, figure 1.1

Examples of specific risk that corresponds to the categories shown in **figure 1.4** include the following:

- Strategic: Changes in customer preference or stakeholder preference, executive turnover
- Environmental: Pollution or disturbance of protected areas
- Market: Foreign-exchange rates, availability of commodities and raw materials, interest rates
- Credit: Callable loans, damage to assets for which the organization is an insurer
- Operational: Employee errors, fraud, theft
- Compliance: Failure to meet regulatory requirements, inaccurate documentation
- IT benefit/value enablement: Delivered projects do not create expected value
- IT program and project delivery: Projects are not delivered in a manner consistent with plans
- IT operations and service delivery: Delivered services fall short of service level agreements (SLAs)

Risk is an influencing factor and must be evaluated at all levels of the enterprises—the strategic level, the business unit level and the information systems level. A properly-managed risk program addresses the impact of risk at all levels and describes how risk at one level may affect the other levels.

IT risk management is the implementation of a risk strategy that reflects the culture, appetite and tolerance levels of risk from organizational management; considers resources, maturity, technology and budgets; and addresses business requirements stemming from regulatory, statutory and contractual compliance drivers. An effective IT risk management strategy is critical to an enterprise's ability to execute its overall business strategy effectively and efficiently.

IT risk management is a cyclical process. The first step in the IT risk-management process is the identification of IT risk, which includes determining the risk context and risk framework, the enterprise's risk appetite and tolerance levels and the process of identifying and documenting risk. The risk identification effort should result in the listing and documentation of the threats posed to organizational assets, efficacy of existing controls, any personnel, processes and technology that may be vulnerable or have a perceived weakness and the potential harm introduced should the threat be realized, which serves as the input for the next phase of the process, IT risk assessment. An assessment effort requires the analysis and evaluation of threats in order to then assess and prioritize risk in context of the enterprise's defined risk appetite and criteria for deviations where risk tolerances would be increased. This provides management with the information needed for risk response and mitigation, the third phase of the cycle, which seeks and implements cost-effective ways to address the risk that has been identified and assessed. The final phase is risk and control monitoring and reporting, in which controls, risk management efforts and the current risk state are monitored, and the results reported back to senior management. The process repeats as the risk environment changes, which may occur as a result of internal or external factors.

Figure 1.5 illustrates the cyclical IT risk-management process.

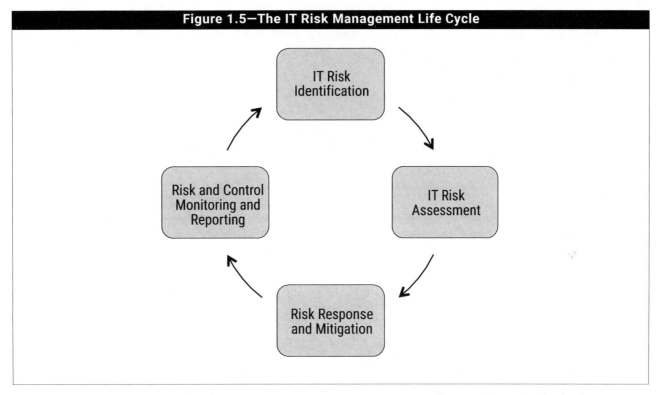

The IT risk-management process is based on the complete cycle of all the elements. A failure to perform any one of the phases in a complete and thorough manner may result in deficiencies being carried forward that cause the overall process to be ineffective. As with all life cycles, the process continues with refinement, adaptation and a focus on continuous improvement and maturity. The more often the risk management life cycle is repeated, the more effective the IT risk management effort will be and the more consistency the organization will see.

1.1.2 Key Concepts of Risk

In order for risk to be discussed within an enterprise, it must be done so meaningfully. All parties must have the same level of understanding of risk. This ensures that when speaking about risk, everyone is speaking the same language. Taxonomies and ontologies for risk must be formally defined, approved, accepted, established and communicated to all parties, ensuring that there is no miscommunication or misunderstanding when discussing risk.

The fundamental nature of risk is that it addresses the odds that some event will happen (likelihood) and what it would mean for the enterprise if that event did happen (impact/consequence). Early attempts to define risk observed that the probability of something happening was a combination of two things: whether something was attempted (threat) and whether the target of the attempt was susceptible to what was tried (vulnerability).

As the study of risk matured, risk practitioners began to distinguish between delineation of the consequences and the extent to which those consequences affected the value-creation activities of the enterprise (impact). It is now common to distinguish between different types of threats, evaluating them on the basis of organizational assets that they may be directed towards, but also unique threat actors that may attempt to exploit any perceived weaknesses or discovered vulnerabilities associated with the asset. This results in the ability to speak about specific threat scenarios in context.

When viewed from the perspective of how these assets are used within the enterprise, it becomes possible to quantify impact in terms of:

- Productivity
- Response costs
- Legal (fines/penalties/summary judgements)
- Competitive advantage
- Damaged reputation/brand impact
- Impaired growth
- Health, safety, security and environment concerns

This is useful, and of added benefit, for two reasons:

- It is easier for managers to set a monetary value of total losses that they are willing to incur (risk appetite) than it is to define what consequences are or are not acceptable in a dozen or more different areas of operations.
- Knowing the potential losses associated with risk provides a basis for deciding how to respond to risk that is beyond acceptable levels because it does not make sense to spend more to respond to a risk than the cost the risk itself presents.

Example of Risk

Consider a house in a dry, wooded area. Wildfire is a threat regardless of the building material used to construct the house (i.e., wood or brick). The likelihood of a wildfire starting in the wooded area during the summer is more probable than during the winter (threat event). This is also distinct from the likelihood of the house burning down (loss event). For the second instance, we take into consideration the building material. A wooden house is more vulnerable to fire; a brick house is not. For the same threat (a fire starting), the likelihood of impact is, therefore, different depending on the vulnerability.

Next, the destruction of the house is a potential consequence. If the house is occupied, the impact is temporary homelessness for those who live there, which imposes the immediate costs of temporary lodging and a replacement wardrobe. Under those circumstances, it makes sense to take precautions sufficient to address this impact, such as insuring the home against fire or putting in a fire-suppression system, but it is not reasonable to hire a full-time fire crew to watch the house on a daily basis because the cost of the fire crew would exceed the cost of the impact. However, if the house were uninhabited and condemned, the consequences would have no negative impact, and no precautions may be necessary.

1.1.3 Importance and Value of IT Risk Management

IT risk management is important to the organization because of the benefits that the program delivers, such as the following:

- Better oversight and understanding of organizational assets
- Reduced or minimized losses
- Identification of threats, vulnerabilities and consequences on a proactive basis
- Prioritization of risk response efforts that align with organizational goals and priorities

- A more holistic basis for and approach to legal and regulatory compliance
- Increased likelihood of project success
- Improved performance
- Greater stakeholder confidence
- Creation of a risk-aware culture with less reliance on specialists
- Better incident, business continuity management, and overall organizational resiliency
- Improved control selection and implementation
- Monitoring and reporting that is meaningful to the organization
- Improved decision making as a result of expanded access to accurate, timely information
- An increased ability to meet business objectives and create value
- Transparency of operations and transactions processing for added value to the enterprise's image

1.1.4 The IT Risk Strategy of the Business

IT risk is the business risk associated with the use, ownership, operation, involvement, influence and adoption of IT within an enterprise. Understanding the business's overall risk strategy should direct the risk practitioner to guide development of an IT risk strategy that aligns with organizational goals and priorities. *All IT risk must be considered both by its impact on IT services and its impact on enterprise operations.* It is this ability to see these dual perspectives which provide the practitioner the appropriate context when analyzing, evaluating, assessing and recommending potential response options.

Many enterprises have some form of risk management in place, but levels of formal definition, documentation and maturity can vary widely between organizations. One role of the risk practitioner is understanding the enterprise's current risk universe, assessing the capabilities and overall maturity of the risk management program. Relevant documentation relating to controls, risk, audit and regulations that influence the organization are available and should be communicated to the appropriate stakeholders. These elements should inform the risk strategy and, if not present, may require the risk practitioner to develop and socialize the necessary documentation for adoption by the enterprise.

Types of IT-related Business Risk

Identifying risk requires a combination of knowledge, experience, imagination and analysis. The risk practitioner is most likely to identify a risk when it takes the form of something that he or she can already envision. For this reason, it is beneficial for the risk practitioner to know the types of IT-related risk faced by the business.

The list of risk types shown in **Figure 1.6** is not exhaustive. The risk practitioner should be committed to continuing education as a professional philosophy and be attentive to the concerns of business-process owners, who may conceive of potential problems specific to their particular areas of expertise that may not be immediately apparent to a generalist. Effective risk management is reliant on a continual dialogue and is a collaborative process between risk practitioner and the appropriate business-process owners and stakeholders.

In addition, the risk practitioner should be aware that IT risk is only one component of the overall risk universe of an enterprise, which also includes risk associated with credit, regulatory compliance, environmental and labor considerations, plus other factors substantially independent of IT. The need to consider the full spectrum of enterprise risk can overwhelm senior managers when individual risk is presented without adequate context. The risk practitioner should always recognize that what is critical to IT may or may not be the most critical consideration of the organization in the context of overall risk.

CHAPTER 1— GOVERNANCE

Figure 1.6—Business-related IT Risk Types	
Type of Risk	Description
Access risk	The risk that information may be divulged or made available to recipients without authorized access by the information owner, reflecting a loss of confidentiality and integrity.
Availability risk	The risk that service may be lost, or data are not accessible when needed.
Cyber and information risk	Failure to ensure the proper safeguarding of privacy, confidentiality, integrity and availability of information, regardless of medium.
Emerging technology risk	Threats associated with the use and implementation of new technology, which have not yet been fully adopted, evaluated or tested for operational resiliency, suitability, sustainability or security.
Infrastructure risk	The risk that the IT infrastructure and systems may be unable to effectively support the current and future needs of the business in an efficient, cost-effective and well-controlled fashion (includes hardware, networks, software, people and processes).
Integrity risk	The risk that data may be unreliable due to incompleteness or inaccuracy.
Investment or expense risk	The risk that the IT investment fails to provide value commensurate with its cost or is otherwise excessive or wasteful, including the overall IT investment portfolio.
Program/project risk	The risk of IT projects failing to meet objectives through lack of accountability and commitment.
Relevance risk	The risk that the right information may not get to the right recipients at the right time to allow the right action to be taken or the right decisions to be made.
Schedule risk	The risk of IT projects taking longer than expected.
Talent risk	The failure to source and the inability to retain the qualified talent needed to achieve the enterprise's strategy.
Third-party risk	Threats introduced by an external entity (vendor, provider, regulatory body) which impact the enterprise.

Senior Management Support

Senior management support is vitally important throughout the risk-management process. With it, the risk-management process is much more likely to have the budget, authority, access to personnel and information, and legitimacy that will provide a successful result. Without it, risk management is almost always unsuccessful. Senior management support should be visible and active, and executives should be willing to intervene when necessary to communicate the importance of risk identification efforts and the need for everyone to actively contribute to the success of the program.

1.1.5 Alignment With Business Goals and Objectives

Effective risk management depends on knowing the enterprise's goals and objectives. The risk practitioner must be careful not to focus solely on a risk from the perspective of one department or one process. The risk practitioner should also consider the risk to other departments, business partners or the overall business goals. It can be easy for IT risk practitioners to calculate risk reflecting the impact only on IT and forget to include the business process-owners that those IT systems support. The risk management program serves the enterprise, so the risk practitioner must remember to look beyond his or her own department.

The best way to understand the goals and objectives of the enterprise is to maintain an active dialogue and regular communication with senior management. The risk practitioner should consult with senior management to validate their understanding of the enterprise's vision and strategy and how those translate down to the appropriate lines of business, use of technologies, growth or potential changes in priorities, keeping in mind that some information may be censored due to ongoing negotiations or confidential strategic plans. Risk considers both the potential gains and possible losses of pursuing new ventures, which may include financial loss, reputational damage, changes in

CHAPTER 1— GOVERNANCE

employment agreements and entry into new markets which introduce new regulatory environments. Knowing what executives are anticipating gives the risk practitioner insight into the various elements that shape the risk universe and plan accordingly as the enterprise evolves.

The underlying importance of risk management is to ensure that risk management is closely aligned with, and integrated into, the strategy, goals and objectives of the enterprise. Because risk management serves the enterprise, executives will ultimately choose a path that seems to offer the best prospects for value creation, and may do so despite the resistance of those involved in risk management. To avoid being seen as obstructionist, the risk practitioner should seek to:

- Understand the business in its proper context
- Listen to and understand the defined strategy
- Proactively seek out ways to secure appropriate technologies and business processes
- Build relationships that promote communication, allowing risk management functions to be incorporated into business processes and new projects
- Be aware of changes to ensure the ability to respond accordingly
- Work to create a culture that encourages open and informed discussions of risk
- Advise on the various aspects of risk, but do not make decisions on behalf of the business

History is a critical component of risk management. The events of the past must be considered in the light of the risk management effort, because an enterprise that has not taken steps to mitigate the future impact of events similar to those that have already occurred may be perceived by stakeholders as irresponsible. However, a narrow focus on the past may result in a lack of vigilance for new threats and future issues, with the enterprise falling into the habit of only fixing events once they have occurred. Alternately, an enterprise that was fortunate in the past may suffer from overconfidence, which can be dangerous and cause it to overlook serious problems that should be thoroughly addressed and mitigated. In this case, the preexisting conditions that led to the event may remain, setting the stage for a serious incident.

1.2 Organizational Structure, Roles and Responsibilities

The effectiveness of the risk management effort is contingent not only on having senior management's support but also is influenced by the positioning of the risk management function within the organizational structure. Ideally, risk management should be a function with enterprise scope, able to reach into all the parts of the organization and provide leadership, advice and direction. An effective risk management program provides timely, accurate and complete information, in a consistent manner, that allows senior management to make informed decisions. To efficiently manage risk, the IT risk must be integrated in the enterprise risk management (ERM) structure, consistent with other departments and business functions.

1.2.1 RACI (Responsible, Accountable, Consulted, Informed)

There are four main types of roles that are involved in the risk-management process:

- **Responsible**: These individuals, often the risk practitioner, are responsible for carrying out risk management efforts. Management can be responsible if risk includes the decisions made by the enterprise.
- **Accountable:** These individuals, often senior management, are accountable for ensuring that risk management efforts are properly staffed, budgeted and are being carried out as planned.
- **Consulted:** These individuals, often business process owners or executive leadership, provide required support and assistance in ensuring that risk management efforts are able to be carried out, in accordance with the direction of senior management.

CHAPTER 1— GOVERNANCE

- **Informed:** These individuals, often senior management or the board of directors, are informed regarding the progress of risk management efforts, as appropriate.

The use of a RACI model can assist in outlining the roles and responsibilities of the various stakeholders. The purpose of a RACI model is to clearly show the relationships between the various stakeholders, the interaction between the stakeholders and the roles that each stakeholder plays in the successful completion of the risk management effort. **Figure 1.7** describes the components of the RACI model. A sample RACI chart is shown in **figure 1.8**.

Figure 1.7—RACI Model	
Assigned Role and Responsibility	Description
Responsible	Individuals tasked with getting the job done, performing the actual work effort to meet stated objectives.
Accountable	The single person liable for the completion of the task, who oversees and manages the person(s) responsible for performing the work effort, who may also play a role in the project. Accountability for a particular task should be assigned to a specific person in order to be effective.
Consulted	Individuals who provide input data, advice, feedback or approvals. Consulted personnel may be from other departments, from all layers of the enterprise, from external sources or from regulators.
Informed	Individuals who are informed of the status, achievement and/or deliverables of the task but who are often not directly responsible for the work effort.

Figure 1.8—Sample RACI Chart				
Task	Senior Management	Steering Committee (Chair)	Department Managers	Risk Practitioner
Collect risk data	I	A	C	R
Deliver the risk report	I	A	I	R
Prioritize risk response	A	I	R	C
Monitor risk	I	A	R	C

Risk management may be applied to an entire enterprise under a singular, centralized, formal risk management team or it may be practiced separately in each level of the enterprise or in regard to specific functions, projects and activities. Functional-level integration is valuable in the context of implementation, but decentralized models can make it unclear to whom the risk practitioner should report within the enterprise. Many enterprises may not have a dedicated chief risk officer (CRO), and of those that do, not all CROs report to the chief executive officer (CEO) or board of directors.

The size and diversity of the enterprise is also a key influencing factor in managing risk. Even in a small company operating in one primary market, where risk management is fairly straightforward, cultural differences between departments can create counterintuitive perspectives. For example, the approach to risk in the finance department may be diametrically opposed to the approach to risk in the sales department. This requires the risk practitioner to work with the management team to develop an understanding of risk and a standard approach to measuring and mitigating risk.

In larger enterprises, risk management may be too big of a task for any one single department or team. This is especially true if the enterprise is geographically dispersed or is responsible for supporting a wide variety of products or services. The risk management effort may need to be arranged by department, products, services or geographic region. In most cases, the organization of the risk management group should follow the same model and the enterprise of the business continuity management team.

1.2.2 Key Roles

Within the business function of risk management there are a number of key roles which should be clearly defined to prevent misunderstandings and miscommunication. The key roles are:

- **Risk manager**—**Responsible** for ensuring that the risk management functions are carried out to support the enterprise's goals and objectives.
- **Risk analyst**—**Responsible** for the analysis, evaluation and assessment of identified threats faced by the enterprise.
- **Risk owner**—The individual in whom the enterprise has invested the authority and **accountability** for making risk-based decisions, and who owns the loss associated with a realized risk scenario.
- **Control owner**—The individual **accountable** for ensuring controls are designed, implemented and operating as planned to keep risk at an acceptable levels. This may also be the risk owner. This includes budgeting, staffing, design, implementation and monitoring of controls.
- **Control stewards**—The individuals who are **responsible** for the routine management and maintenance of controls on behalf of the control owner and institute changes at the direction of the control owner.
- **Subject matter experts**—These are individuals who have an intimate knowledge and can provide valuable insight into specific areas (e.g., financial, IT, call center) within an organization as they relate to identified or perceived threats and risk to the enterprise; based upon their experience, understanding and expertise in their respective domain.

1.2.3 Organizational Structure and Culture

The structure and culture of the organization are critical success factors to the risk-management program, as they directly influence and inform staff decisions relating to risk prevention, risk detection and risk response efforts. A mature enterprise has policies and procedures and an effective reporting and notification structure in place to detect, notify and escalate a situation effectively. For instance, an enterprise that does not have a mature incident response capability will often react to incidents in an unpredictable, *ad hoc* reactive manner and can expect to experience inconsistent results.

The risk management function should have an enterprise mandate that allows risk practitioners to review and provide input into all business processes, participate in incident management activities and be responsible for reviewing incidents for lessons learned and to improve incident response planning, detection and recovery. Lessons learned in one department or on one system or application may be applicable in protecting other departments, systems or applications from the same problems, so collaboration and sharing of information are important parts of using risk scenarios.

The enterprise's behavior and attitude towards risk is typically personified in its approach and view of matters relating to risk. If the organizational culture is to hide problems rather than communicate or address them—or to only use an adverse situation to point blame—then the ability of the risk practitioner to effectively contribute to the protection of the enterprise and assist in the investigation of an incident may be severely impaired. **Figure 1.9** illustrates a sample organizational structure.

CHAPTER 1— GOVERNANCE

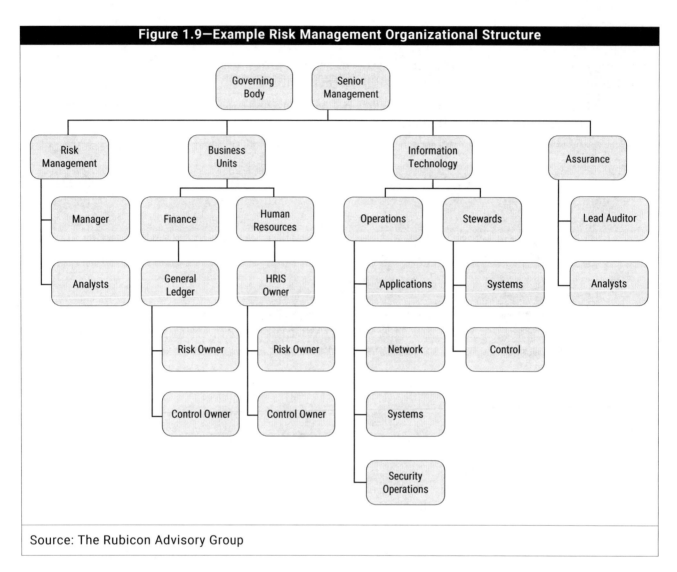

Figure 1.9—Example Risk Management Organizational Structure

Source: The Rubicon Advisory Group

1.3 Organizational Culture

Risk management is a core part of corporate governance. The governance of the assets and mission of the enterprise is reflected in the ways in which the enterprise seeks to protect its assets and attain its goals, and risk is a factor that may lead to failure or a loss of asset value. Understanding risk includes understanding the goals, objectives, values and ethics of the enterprise.

These help determine the organizational culture of the enterprise. **Figure 1.10** describes some common organizational cultures related to risk.

CHAPTER 1— GOVERNANCE

Figure 1.10—Organizational Culture Relating to Risk

Vulnerable	Reactive	Compliant	Proactive	Resilient
Don't Care Culture	**Blame Culture**	**Compliance Culture**	**Ownership Culture**	**Way of life**
• Apathy • Near misses not considered • Negligence • Hiding of incidents • No or little training • Poor or no communication	• Resistance to caring • Some near-miss reporting • Some "window dressing" • Ad hoc/inconsistent training • Communication on a need-to-know basis	• Responsibilities assigned • Reporting limited to compliance areas • "As required" process definition • Limited instrumentation and investments • Minimal required training • Compartmentalized communications	• Clear lines of accountability and responsibility defined • Processes defined to enhance long-term sustainability and operationalization • Appropriate instrumentation and investments are made • Training defined and required • Open communication	• Lines of accountability and responsibility are communicated and understood throughout the enterprise • Active monitoring and reporting • Advanced instrumentation and investments made for benefits of improvements and optimization • Training is encouraged • Active communication
Accept that incidents happen	*Prevent a similar incident*	*Prevent incidents before they occur*	*Continuous improvement of systems*	*Way we do business*
Reactive approach	Administrator-driven	Compliance-driven	Business-driven	Risk-driven
• No process defined • Legal noncompliance • Accept process decay • Superficial incident investigations • No risk assessment • No monitoring/audits • Permit noncompliance • Potential illegal activity	• Reactive risk assessment • Minimum legal compliance • Incident investigation, but limited analysis • Focus on what happened • No systems focus • Human fault focus • No network hygiene or management • Ad hoc monitoring/audits	• HIPAA, SOX, PCI-driven • Risk assessment through existing systems • Legal compliance • Planned network hygiene and management initiatives • Periodic testing and evaluations • Casual incident analysis based on even potential • Planned monitoring/audits	• Formal risk assessment • Beyond legal compliance • Seek to actively engineer out process/system inadequacies • Incident lessons learned shared at all levels • Well-designed plans, processes and procedures • Focus on adhering to plans and procedures • Integrated audit	• Individually internalized • Integrated management system • Risk assessment integrated at all levels • Self-regulating • Reduce/eliminate problems before they occur • All threats considered in decision making • Enhancement through evaluation/audit

1.3.1 Organizational Culture and Behavior and the Impact on Risk Management

One of the first challenges faced by a risk practitioner is to determine the risk appetite of senior management. Examples of risk that management must evaluate and set criteria for accepting include deciding whether the enterprise should:

- Invest
- Take on a new line of business

CHAPTER 1— GOVERNANCE

- Develop a new product
- Open a new office
- Hire a new employee
- Invest in new hardware or software
- Upgrade existing applications
- Implement new controls

The risk practitioner should be aware that the risk appetite may change over time or vary based on the type of risk. Thus, the risk appetite requires periodic review. Depending on market conditions, confidence, past successes or failures, global economics, reports in the media, availability of resources, new regulations or long-term strategy, the extent to which risk appetite changes may be dramatic. Culture and ethics also play a role in determining variances from the defined risk appetite.

Variations from the defined risk appetite will also occur and should be based on a predefined set of criteria to establish the appropriate risk tolerances. The need for these criteria to be established prior to risk management efforts ensures that responses to risk are considered in the proper context.

For one example, a business unit may establish criteria prior to kicking off a project that allows for an increase of up to 20 percent in the project's duration, while at the same time only allowing for a 10 percent increase in costs.

Another example is a financial institution that has generally defined a low appetite for credit card fraud and has implemented aggressive anti-fraud safeguards and countermeasures. During certain seasonal periods, anti-fraud capabilities may be relaxed as consumer purchases increase. This decision is made as it anticipates that the revenue generated during the season will exceed losses from fraudulent activity, losses incurred due to consumers' displeasure in having to respond to normal anti-fraud controls and the short duration of the seasonal period (30 days).

Culture

A culture drives the behaviors of personnel, and people will often act according to their environment. Risk culture is defined as the set of shared values and beliefs that governs attitudes toward risk taking, care and integrity and determines how openly risk and losses are reported and discussed. All employees of the enterprise should be aware of the risk culture, because knowing the risk culture provides a useful context for their own decisions and actions.

A person will often act according to his or her beliefs, so creating an environment that considers people's belief systems is often an effective method of changing behaviors. For instance, a culture of honesty and openness may reduce the risk of theft, inappropriate actions or attacks as employees may be less frustrated and more engaged in the risk management program.

The risk practitioner should be aware that departments may have their own subcultures that differ from the organizational culture. Adherence to a subculture may lead to people taking more risk than management actually intends to accept; it may also manifest in people refusing to take risk when management would like them to do so. Subcultures may evolve out of industry norms, but they can also arise organically as a reaction to demonstrated actions. For instance, in an enterprise in which senior executives encourage employees to take risks but blame them when something goes wrong, creativity and innovation may be less pronounced than would otherwise be the case.

Risk Awareness

Awareness is a powerful tool in creating the culture, forming ethics and influencing the behavior of the members of an enterprise. The staff and operational teams are often the first to be aware of any problems or abnormal activities. Through awareness programs, it is possible to develop a team approach to risk management that enables every member to identify and report on risk and to work to defend systems and networks from attacks. They can help identify vulnerabilities, suspicious activity and possible attacks. This may enable a faster response and better containment of a risk when an attack happens.

Risk awareness acknowledges that risk is an integral part of the business. This does not imply that all risk is to be avoided or eliminated, but rather that:

- Risk is well understood and known.
- IT risk issues are identifiable.
- The enterprise recognizes and uses the means to manage risk.

A risk awareness program creates an understanding of risk, risk factors and the various types of risk that an enterprise faces. An awareness program should be tailored to the needs of the individual groups within an enterprise and deliver content suitable for that group. A risk awareness program should not disclose vulnerabilities or ongoing investigations except where the problem has already been addressed. Using examples and descriptions of the types of attacks and compromises that other organizations have experienced can reinforce the need for diligence and caution when addressing risk.

Awareness education and training can serve to mitigate some of the biggest organizational risk and achieve the most cost-effective improvement in risk and security. This can generally be achieved by educating an organization's staff in required procedures and policy compliance, along with ensuring that staff can identify and understand the risk that threatens the organization. It is critical that the training effectively communicate the risk and its potential impact in order for staff to understand the justification for what may be seen as inconvenient extra steps that risk mitigation and security controls often require.

The risk practitioner must also understand the enterprise's structure and culture and the types of communication that are most effective to develop awareness and training programs that will be effective. Periodically changing risk awareness messages and the means of delivery will help maintain a higher level of risk awareness. Procedural controls can be complex, and it is essential to provide training as needed to ensure that staff understands the procedures and can correctly perform the required steps.

Awareness of information security policies, standards and procedures by all personnel is essential to achieving effective risk management. Employees cannot be expected to comply with policies or standards that they are not aware of, or follow procedures they do not understand. The risk practitioner should advise for a standardized approach, such as short computer- or paper-based quizzes to gauge awareness levels. Periodic use of a standardized testing approach provides metrics for awareness trends and training effectiveness. Further training needs can be determined by a skills assessment or employing a testing approach. Indicators for additional training requirements can come from various sources such as tracking help desk activity, operational errors, security events and audits.

An awareness program for management should highlight the need for management to play a supervisory role in protecting systems and applications from attack. A manager has the responsibility to oversee the actions of his or her staff and direct compliance with the policies and practices of the enterprise.

Awareness training for senior management should highlight the liability, need for compliance, due care and due diligence and the need to create the tone and culture of the enterprise through policy and good practice. Senior management may need to be reminded that they are the ones who "own" the risk and bear the responsibility for determining risk acceptance levels.

1.3.2 Risk Culture

Consciously or unconsciously, senior management develops and conveys a particular level of willingness to embrace, cautiously accept or avoid risk; this is called the risk culture of the enterprise. In order to successfully embed an effective risk culture within an enterprise, senior management needs to lead from the top, supporting these concepts not only with their words, but also with their actions.

The best indicator of the enterprise's risk culture is how it makes decisions on how to best respond to various risk. Risk culture reflects a balance between weighing the negative, positive and regulatory elements of risk, being informed as to the direction and response relating towards taking risk, as seen in **figures 1.12** and **1.13**. Symptoms of an inadequate or problematic risk culture are listed in **figure 1.14**.

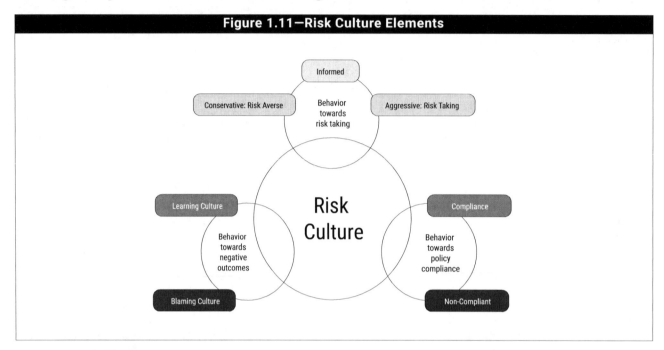

Figure 1.12—Description of Risk Culture Elements	
Element	Description
Behavior toward taking risk	How much risk does the enterprise believe that it can absorb, and what specific risk is it willing to take?
Behavior toward policy compliance	To what extent do people within the enterprise embrace or comply with policy?
Behavior toward negative outcomes	How does the enterprise deal with losses, missed opportunities, and other negative outcomes? Does it learn from them and try to adjust, or does it assign blame without treating the root cause?

Figure 1.13—Symptoms of an Inadequate or Problematic Risk Culture	
Symptom	Description
Misalignment between the risk appetite stated in policy and reflected in behavior	Management promulgates policies that mandate an approach to risk that is largely or completely different from demonstrable behavior by and within the enterprise.
Existence of a "blame culture"	Discussions focus on symptoms and accountability for problems. Root causes are rarely identified.
	Business units tend to blame IT when projects are not delivered on time or do not meet expectations. In doing so, they fail to realize how the business unit's involvement upfront affects project success.
	In extreme cases, the business unit may assign blame for a failure to meet the expectations that the unit never clearly communicated. The "blame game" only detracts from effective communication across units, further fueling delays. Executive leadership must identify and quickly control a blame culture if collaboration is to be fostered throughout the enterprise.

1.3.3 A Risk-driven Business Approach

A mature organization is one that is able to make informed decisions relating to the various risk, in its respective business context. This means that the risk practitioner is able to carry out risk management efforts that aligns with the defined risk appetite. The risk practitioner is cognizant of the criteria associated with variances to the established and communicated risk appetite. This ability ensures that only the risk that can be accepted are recorded and monitored moving forward, and that efforts are spent on areas where the risk has exceeded the established appetite and tolerance levels. **Figure 1.14** describes a sample approach for assessing risk associated with data loss.

CHAPTER 1— GOVERNANCE

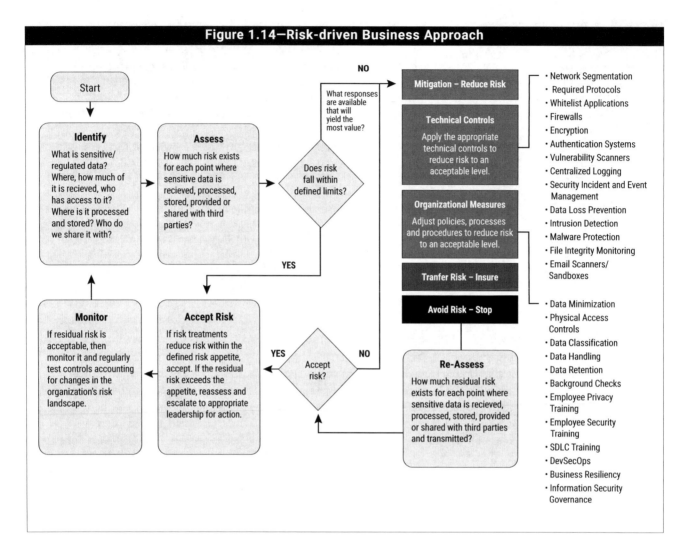

Figure 1.14—Risk-driven Business Approach

1.3.4 The Value of Risk Communication

The discipline of risk management is related to audit, business continuity management and security. Having a reporting relationship that facilitates communication regarding risk, threats, incidents, vulnerabilities and assets among these groups will result in a greater level of accuracy for the risk-management process.

Communication of the threats faced by the enterprise, incidents experienced, existing vulnerabilities and the business value of assets is an essential part of the risk-management process. Not every risk incident needs to be reported, but a lack of communication can be a sign that an enterprise is not healthy or stable. If managers are not open and transparent with regard to issues and operational failures, decisions may be made on the basis of incomplete or inaccurate information. The risk practitioner should seek to develop lines of communication and reporting so that information is available to management on a timely basis and encourage communication of even negative activities when appropriate.

The method and openness of risk communications play a key role in defining and understanding the risk culture of the enterprise. Communication is important because it removes the uncertainty and doubts concerning risk management and supporting efforts. If risk is to be effectively and efficiently managed and mitigated, it must first be openly discussed, properly communicated, and understood by the various stakeholders and personnel throughout the enterprise in ways that are appropriate for their respective roles. The benefits of open communication of risk include:

CHAPTER 1— GOVERNANCE

- More informed risk decisions by the appropriate stakeholders due to an improved understanding of actual exposure and potential business impact
- Greater awareness among all stakeholders of the importance and value of integrating risk management into their daily duties
- Transparency to external stakeholders regarding both the actual levels of risk facing the enterprise and the risk-management processes in use

The consequences of poor communication on risk include:

- An overconfidence in the enterprise's capabilities and controls pertaining to the risk universe
- A false sense of security at all levels of the enterprise
- Improper categorization of risk
- Misallocation of resources
- Reduced value to the stakeholders

Unbalanced or ineffective communication of an organization's risk to external stakeholders has the potential to lead to incorrect and negative perceptions by third parties, such as:

- Clients
- Investors
- Regulators

Ineffective communication can also give the perception that the organization is trying to hide risk from stakeholders.

Figure 1.15 describes the broad array of information flows and major types of IT risk information that should be communicated.

Figure 1.15—Risk Components to Be Communicated	
Type	Description
Expectations from risk management	• Risk strategy, policies, procedures, awareness training and continuous reinforcement of principles. This is essential communication regarding the enterprise's overall strategy toward IT risk that: • Drives all subsequent efforts on risk management • Sets the overall expectations about the risk management program
Current risk management capability	• Allows for monitoring of the state of the "risk management engine" in the enterprise • Is a key indicator for good risk management • Has predictive value for how well the enterprise is managing risk and reducing exposure
Status (actual, regarding IT risk)	• The risk profile of the enterprise (i.e., the overall portfolio of [identified] risk to which the enterprise is exposed) • Key risk indicators to support management reporting on risk • Event/loss data • The root cause of loss events • Options to mitigate risk (including cost and benefits)

CRISC® Review Manual 7th Edition
ISACA. All Rights Reserved.

CHAPTER 1— GOVERNANCE

1.4 Policies and Standards

Policies provide direction regarding acceptable and unacceptable behaviors and actions to the enterprise. Standards and procedures support the requirements defined in the policies set forth by the enterprise's senior management.

1.4.1 Policies

Policies empower risk management, audit and security staff. Policies should clearly state the position of senior management toward the protection of information, which will allow the development of procedures, standards and baselines that reflect management priorities. Executive sponsorship also provides a mandate that all departments comply with policy requirements.

Enterprises often have several layers of policy in order to allow delegation of authority. A high-level policy may be issued by senior management as a way to address the objectives of the enterprise's mission and vision statement. This overarching policy does not have a technical focus in order to prevent it from becoming outdated when technology changes. High-level policy may direct compliance with laws and good practices, and is likely to state the goal of managing risk through protecting the enterprise's assets, including the information and information systems that support business operations.

High-level policies are instrumental in determining the approach of the enterprise toward risk management and acceptable levels of risk. Without policies in place, the risk practitioner may not be able to gain access to key personnel, be left out of strategic planning sessions or be ignored when performing investigations. Below the high-level policies are technical and functional policies that include specifics regarding technology use, such as remote access, acceptable use and passwords. These policies are subject to change as technology changes and new systems are developed.

The risk practitioner should identify the presence or lack of policies and work to determine whether the policies are enforced. If policies are not developed and communicated, the enterprise has no means of enforcing standards of behavior, which increases the risk that the behavior may be inappropriate. A lack of enforcement may also lead to circumvention of controls or increased liability because the enterprise has admitted the need for a policy but does not follow its own rules. The risk practitioner should assess the risk associated with the policy framework of the enterprise and provide recommendations as necessary. Where policies are out of date, lack enforcement or are incomplete, the risk practitioner should clearly document the vulnerability and associated risk that this poses to the enterprise.

Specific to both governing the risk-management program and performing risk-management activities, **Figure 1.16** lists policies that should be implemented within an enterprise.

Figure 1.16—Risk Management Policies

Policy	Description
Enterprise risk policy	Defines governance and management of enterprise risk at strategic, tactical and operational levels, pursuant to business objectives. Translates enterprise governance into risk governance principles and policy and elaborates risk management activities.
Information security policy	Sets behavioral guidelines relating to the protection of corporate information, to include the underlying infrastructure and supporting systems.
Privacy policy	Sets behavior guidelines to protect information deemed personally identifiable or non-public as defined by appropriate statutory drivers.
Risk appetite/tolerance policy	Clearly defines the amount of risk the enterprise is willing to accept in pursuit of goals and objectives, in addition to criteria that may allow, with appropriate stakeholder approval, for a temporary increase of risk to the enterprise.
Risk acceptance policy	Clearly identifies who has the authority to accept risk on behalf of the enterprise and to what level of risk they are authorized to accept.

1.4.2 Standards

A standard is a mandatory requirement, code of practice or specification approved by a recognized external standards organization, such as ISO. Standards are implemented to comply with the requirements and direction of policy to limit risk and support efficient business operations.

Many enterprises realize that the value of standards is the authority and proven value that they provide. A standard mandates the way in which personnel in an organization must comply with recognized practices or specifications and can help ensure a consistent approach to meeting risk requirements across an organization. An enterprise may base its practices and operations on external standards, such as an ISO standard, or it may develop and tailor its own standards, such as requiring all staff to use the same product, operating system or desktop. The use of a standard facilitates support and maintenance, provides better cost control and provides authority for the practices and procedures of the organization because a standard requires the implementation of certain practices.

Common standards are listed in **figure 1.17**.

Figure 1.17—Common Risk Standards

Standard	Organization
A Risk Management Standard	United Kingdom Institute of Risk Management
Enterprise Risk Management	The Committee of Sponsoring Organizations
AS/NZS Standard 4360	Standards Australia/Standards New Zealand
BSI 31100 Risk Management Code of Practice	British Standards Organization
ISO/IEC 31000 Risk Management Principles and Guidelines	International Organization for Standardization
Managing Information Security Risk	National Institute of Standards & Technology
Factor Analysis of Information Risk	The FAIR Institute
SABSA Risk Assurance and Guidance	The SABSA Institute

1.4.3 Procedures

Procedures are more granular than standards and support their implementation. A procedure is a document containing a detailed description of the steps necessary to perform specific operations in conformance with applicable standards. Procedures are defined as part of processes and created in order to define the ways in which the processes should be carried out. Procedures are invaluable as a means of implementing the intent of policy, because they define the tasks that people actually carry out. By defining actions in consistent and measurable ways, the enterprise can greatly

CHAPTER 1— GOVERNANCE

increase the probability that an operation is conducted according to good practice. Therefore, the risk practitioner can be assured that any abnormal operations will be detected.

A lack of standards and procedures makes it difficult to carry out activities in a systematic manner and may result in undependable, inconsistent operations and elevated risk. The risk practitioner should also distinguish between the existence of published procedures and their actual use. It is common for procedures to be followed only for a short time, after which point experienced staff begin to work from memory. This practice should be discouraged in any environment in which precision is important, such as shutdown procedures for power plants or industrial machinery or complex monetary transactions.

Specific to risk management efforts, **Figure 1.18** lists common risk management procedures.

Figure 1.18—Risk Management Procedures	
Procedure	Description
Risk identification	Details the steps associated with how threats and risk is identified in the enterprise and how they are to be cataloged and reported to risk management staff for subsequent analysis, evaluation and assessment.
Risk analysis	Details the steps associated with how risk analysis will be performed; includes a taxonomy and ontology associated with risk.
Risk evaluation	Details the steps associated with how risk is evaluated by the enterprise, establishing business context, sufficient detail and the criteria that will allow for the analyst to understand the risk in the appropriate context.
Risk assessment	Details the steps associated with how the enterprise assesses the business impact a risk poses to the ability to achieve their goals and objectives.
Risk response	Details the steps associated with how to respond to risk, specifically the criteria associated with the appropriate options of accept, avoid, mitigate or transfer/share, in context of the enterprise's defined risk appetite, goals and objectives.
Control selection	Details the steps associated with how controls (safeguards or countermeasures) are selected when responding to risk. This should consider the business value of the asset, threat type and event associated with the risk that the control would be addressing to ensure the most effective and efficient control is selected.
Control monitoring	Details the steps associated with how a control is monitored and who is responsible for ensuring that the control is operating as designed and implemented. Control monitoring ensures that risk is maintained at the appropriate level and that any variances regarding operational effectiveness, management and performance are consistently reviewed to reduce the likelihood of the re-introduction of unacceptable risk to the enterprise.
Establishing key performance indicators (KPIs)	Details the steps associated with how KPIs are determined by the organization. This should include the thresholds, tolerances and baselines, and what information is required by the business owner who is responsible for the risk.
Establishing key control indicators (KCIs)	Details the steps associated with how KCIs are determined and defined by the enterprise; KCIs quantify how effectively a specific control tool, approach or methodology is working.
Establishing key risk indicators (KRIs)	Details the steps associated with how KRIs are determined and defined by the enterprise; KRIs monitor risk to ensure that any changes to the risk profile are maintained within the appropriate levels.
Risk monitoring	Details the steps associated with how risk will be monitored by an enterprise; given the ever-changing nature of the risk universe. Considerations should be given not only for the risk that is actively being addressed through mitigation, but also the risk that has been accepted, to ensure it is still within the enterprise's defined acceptable levels.
Risk reporting	Details the steps associated with what frequency, level of detail and escalation criteria relating to how risk is reported to the appropriate stakeholders.

1.4.4 Exception Management

Having policies, standards and procedures in place is an essential part of operating secure systems and attaining a secure state; however, there may be cases in which an exception to policy, standard or procedure is necessary. If exceptions are undocumented and uncontrolled, the level of risk is unknown and may result in an undesired level of risk or overconfidence in the effectiveness of established controls. Therefore, exceptions should only be allowed through a documented, formal process that requires approval of the exception by a senior manager. The risk practitioner must ensure that an exception management process is in place and is being followed, and that exceptions are removed when they are no longer needed.

1.4.5 Risk Management Standards and Frameworks

Several good sources for risk identification and classification standards and frameworks are available to the risk practitioner. The following list is not comprehensive, and many more standards are available. However, this list may allow the risk practitioner to consider a framework or standard that would be suitable for use in their enterprise. Many countries and industries have specific standards that must be used by organizations operating in their jurisdiction. The use of a recognized standard may provide credibility and completeness for the risk assessment and management program of the enterprise and help ensure that the risk management program is comprehensive and thorough.

Note: The CRISC candidate will not be tested on any specific standard. The use of any standards in this review manual is for example and explanatory purposes only.

ISO 31000:2018 Risk Management — Guidelines

ISO 31000:2018 states[1]:

> *This document is for use by people who create and protect value in organizations by managing risk, making decisions, setting and achieving objectives and improving performance. Organizations of all types and sizes face external and internal factors and influences that make it uncertain whether they will achieve their objectives.*
> *Managing risk is iterative and assists organizations in setting strategy, achieving objectives and making informed decisions.*
> *Managing risk is part of governance and leadership and is fundamental to how the organization is managed at all levels. It contributes to the improvement of management systems.*
> *Managing risk is part of all activities associated with an organization and includes interaction with stakeholders.*
> *Managing risk considers the external and internal context of the organization, including human behavior and cultural factors.*

ISO 31000:2018 also describes three major components for risk management (**Figure 1.20**).

CHAPTER 1—GOVERNANCE

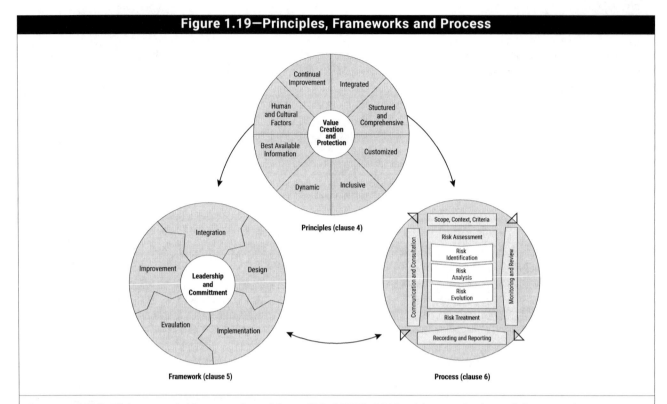

Figure 1.19—Principles, Frameworks and Process

Source: ©ISO. This material is reproduced from ISO 31000:2018 with permission of the American National Standards Institute (AnSI) on behalf of ISO. All rights reserved.

COBIT and Information Risk

COBIT is a framework for the governance and management of information and technology, aimed at the whole enterprise. Enterprise information and technology includes all the technology and information processing the enterprise puts in place to achieve its goals, regardless of where this happens in the enterprise. In other words, enterprise information and technology is not limited to the IT department of an organization but certainly includes it.

The *COBIT Focus Area: Information Risk* provides specific guidance related to information and technology risk based on COBIT core guidance.

IEC 31010:2019 Risk Management—Risk Assessment Techniques

IEC 31010:2019 states:[2]

> *Organizations of all types and sizes face a range of risk that may affect the achievement of their objectives. These objectives may relate to a range of the organization's activities, from strategic initiatives to its operations, processes and projects, and be reflected in terms of societal environmental, technological, safety and security outcomes, commercial, financial and economic measures, as well as social, cultural, political and reputation impacts.*
> *All activities of an organization involve risk that should be managed. The risk-management process aids decision making by taking account of uncertainty and the possibility of future events or circumstances (intended or unintended) and their effects on agreed objectives.*

ISO/IEC 27001:2013 Information Technology—Security Techniques—Information Security Management Systems—Requirements

ISO 27001:2013 states:[3]

> The organization shall define and apply an information security risk assessment process that:[...]
> c) identifies the information security risk: apply the information security risk assessment process to identify risk associated with the loss of confidentiality, integrity and availability for information within the scope of the information security management system; and identify risk owners.

ISO/IEC 27005:2018 Information Technology—Security Techniques—Information Security Risk Management

ISO/IEC 27005 states:[4]

> This international standard provides guidelines for information security risk management in an organization, supporting in particular the requirements of an information security management system (ISMS) according to ISO/IEC 27001. However, this standard does not provide any specific methodology for information security risk management. It is up to the organization to define their approach to risk management, depending for example on the scope of the ISMS, context of risk management, or industry sector. A number of existing methodologies can be used under the framework described in this International standard to implement the requirements of an ISMS.

NIST Special Publications

NIST has a wide range of special publications available at *csrc.nist.gov*. Publications related to IT risk are discussed in the following sections.

NIST Special Publication 800-30 Revision 1: Guide for Conducting Risk Assessments

NIST Special Publication 800-30 Revision 1 describes risk assessment in the following manner:[5]

> Risk assessments are a key part of effective risk management and facilitate decision making at all three tiers in the risk-management hierarchy including the organization level, mission/business process level and information system level. Because risk management is ongoing, risk assessments are conducted throughout the system development life cycle, from pre-system acquisition (i.e., material solution analysis and technology development), through system acquisition (i.e., engineering/manufacturing development and production/deployment), and on into sustainment (i.e., operations/support).

NIST Special Publication 800-39: Managing Information Security Risk

NIST Special Publication 800-39 states:[6]

> The purpose of Special Publication 800-39 is to provide guidance for an integrated, organization-wide program for managing information security risk to organizational operations (i.e., mission, functions, image and reputation), organizational assets, individuals, other organizations and the Nation resulting from the operation and use of federal information systems. Special Publication 800-39 provides a structured, yet flexible approach for managing risk that is intentionally broad-based, with the specific details of assessing, responding to and monitoring risk on an ongoing basis provided by other supporting NIST security standards and guidelines.

Example of a Risk Management Program Based on ISO/IEC 27005

Especially in cases where good risk management practices are not currently in place, the risk practitioner may wish to formally adopt or informally use one or more standard or framework to guide his/her risk management program. The program encompasses the goals of the organization as a whole, not specifically the roles and duties of the risk practitioner.

Risk management includes several components, and although the specific names of each component may differ from one standard or framework to the next, the typical progression is to first identify areas of risk and analyze them, then use the results of these analyses to choose appropriate ways of responding to them, after which the organization monitors the effectiveness of its implemented choices and reports progress to senior management. **Figure 1.20** illustrates one such program based on ISO/IEC 27005:2011.

Note: The following is provided as an example of a risk-management process. The risk practitioner may find that they will need to adopt one or more well-established standards or frameworks to ensure that the risk management program in their organization is complete and authoritative.

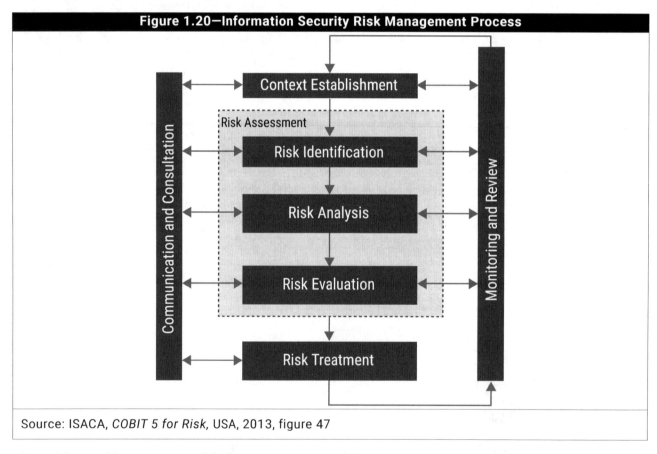

Source: ISACA, *COBIT 5 for Risk*, USA, 2013, figure 47

Figure 1.21 provides a summary of the important concepts of these process steps.

CHAPTER 1— GOVERNANCE

Figure 1.21—ISO/IEC 27005 Process Steps	
ISO/IEC 27005 Process Step	Important Concepts of the Component
Context Establishment	This process step includes: • Setting the basic criteria necessary for establishment of information security risk management (ISRM) • Defining the scope and boundaries • Establishing an appropriate organization operating the ISRM
Risk Assessment	Risk assessment determines the value of the information assets, identifies the applicable threats and vulnerabilities that exist (or could exist), identifies the existing controls and their effect on the risk identified, determines the potential consequences and prioritizes the derived risk and ranks it against the risk evaluation criteria set in the context establishment. This process step consists of risk identification, risk analysis and risk evaluation.
Risk Identification	Risk identification includes the identification of: • Assets • Threats • Vulnerabilities • Existing controls • Consequences The output of this process is a list of risk scenarios with their consequences related to assets and business processes.
Risk Analysis	The risk analysis step includes: • Assessment of consequences • Assessment of incident likelihoods • Determination of level of risk
Risk Evaluation	In this step, levels of risk are compared according to risk evaluation criteria and risk acceptance criteria. The output is a prioritized list of risk elements and the incident scenarios that lead to the identified risk elements.
Risk Treatment	Risk treatment options include: • Risk mitigation • Risk acceptance • Risk avoidance • Risk transfer/sharing
Risk Acceptance	The input is a risk treatment plan and the residual risk assessment subject to the risk acceptance criteria. This stage comprises the formal acceptance and recording of the suggested risk treatment plans and residual risk assessment by management, with justification for those that do not meet the enterprise's criteria.
Risk Communication and Consultation	This is a transversal process where information about risk should be exchanged and shared between the decision maker and other stakeholders through all the steps of the risk management process.
Risk Monitoring and Review	Risk and its influencing factors should be monitored and reviewed to identify any changes in the context of the organization at an early stage and to maintain an overview of the complete risk picture.

1.5 Business Process Review

A business process review examines the effectiveness and efficiency of an enterprise in meeting its goals and objectives. When responding to risk, the organization should review its business processes to ensure that the correct solution is selected to integrate and work effectively with the existing environment, or whether the environment should be reengineered to improve its efficiency and success in meeting organizational objectives.

A business process review requires input from knowledgeable representatives from all affected departments within the organization, and it may also bring in external experts who can provide advice and assistance. The purpose of the business process review is to identify ways to:

- Identify problems or issues with the current process
- Gather information toward improving processes
- Prepare a road map to implement required changes
- Assign responsibility and accountability for projects
- Schedule individual projects according to priority
- Monitor project progress for attainment of milestones and production of deliverables
- Review and obtain feedback on project results
- Verify compliance to standards and policies

The steps of a business process review are:

1. Document and evaluate current business processes
    a. List critical processes, supply chains and services
        i. Classify critical processes by granularity level (e.g., first, second or third degree)
        ii. Identify responsibility and accountability for each process
    b. Document current business processes and risk
        i. Review documentation
        ii. Interview management, users and other stakeholders
        iii. Observe actual processes
        iv. Validate with business representatives
            A. Validate training and skills required of personnel
    c. Document identified issues and problems
    d. Baseline other organizations
    e. Work with team members to discover potential solutions and improvements
2. Identify potential changes
    a. Use focus groups and workshops to determine process improvements
    b. Validate proposed changes with management and obtain approval to proceed
3. Schedule and implement changes
    a. Design changes
    b. Identify dependencies
    c. Communicate the schedule of changes
4. Feedback and evaluation
    a. Measure operational efficiencies
        i. Customer satisfaction
        ii. User satisfaction
        iii. Improvements in productivity/quality
        iv. Feedback to improve processes

Inefficient or outdated business processes may pose a risk by making enterprises noncompetitive. Business processes should be flexible enough to adapt to changes in the market or technology. Periodic reviews of critical business processes should be performed to allow for updating as appropriate. They should also assess the relevancy, benefit and value changes and modifications as appropriate.

1.5.1 Risk Management Principles, Processes and Controls

Risk management exists to support the enterprises, so implementation should not unduly affect business operations. The risk practitioner should consider the impact of processes and controls on the ability of the business to meet its objectives and of users to accomplish their tasks in a simple, logical manner. The risk practitioner should present this information to the risk owner in a way that allows them to weigh these considerations against the consequences associated with the risk.

Principles of Risk Management

While there is a strong reliance on information and related technologies used within the enterprise, risk originating from IT, information security and cybersecurity is not the only risk that requires the attention of risk practitioners. Risk originating from various lines of business exists—such as risk associated with process failures, trends that impact forecast business or economic cycles or acts of nature—also need to be considered and ultimately managed. Any threat that has the potential for preventing the enterprise from meeting their defined business goals and objectives should be managed from the perspective of overall enterprise objectives, and thus falls under the guiding principles of the Risk IT Framework (**figure 1.22**):

- Connect management of IT-related risk to business or mission objectives.
- Align the management of IT-related business or mission risk with ERM when possible.
- Balance the costs and benefits of managing IT-related risk.
- Promote ethical and open communication of all IT-related risk.

Establish the tone at the top while defining and enforcing personal accountability for operating within acceptable and well-defined tolerance levels. Use a consistent approach, integrated into daily activities, that is standard, repeatable and aligned to strategy.

CHAPTER 1— GOVERNANCE

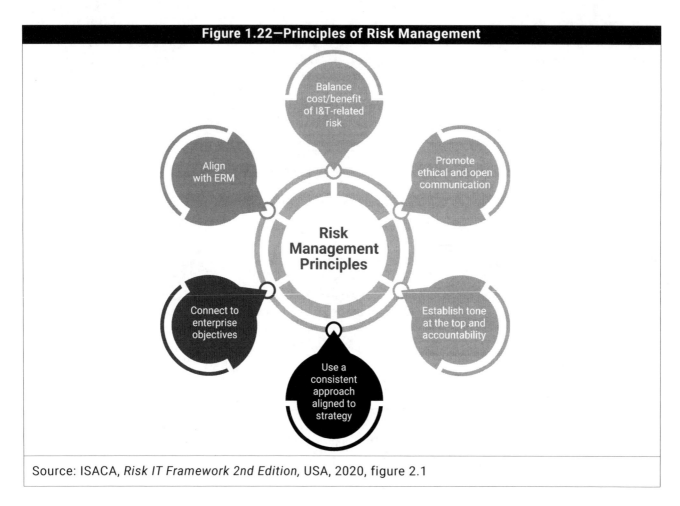

Figure 1.22—Principles of Risk Management

Source: ISACA, *Risk IT Framework 2nd Edition*, USA, 2020, figure 2.1

Processes and Controls

Whether implementing new or modifying existing processes and controls, care and consideration must be given to how they will integrate into the existing environment. The introduction of change within any environment can be met with resistance, so the benefits and values have to be properly socialized and communicated with the appropriate stakeholders. Often, the introduction or modification of processes or controls is viewed as an inhibitor to the workforce in that the process or control will add another step or prevent the workforce from performing their activities. For this reason, when possible and especially with the introduction of new technologies, they should be as transparent to the end-user as possible. The introduction of transparent technical controls limit the likelihood of a member of the workforce from finding a "workaround" to the control, leaving the enterprise with a false sense of security; but also reduces the view that the implementation of the control, and by extension the risk practitioner, is preventing the successful completion of their activities.

Also, the risk practitioner needs to highlight the benefit and value gained through the introduction or modification of processes or controls, and how through their adoption it demonstrates alignment with the various regulations and standards. Consider a business process that involves the processing and storage of sensitive information, such as personal health information (PHI): protecting the information from unauthorized disclosure, alteration or deletion is of high importance. Sensitive data may be masked or hidden from users who do not have a need to know or may be read-only for users who do not have authority to alter the data. Building these types of controls into systems and applications can be an effective form of preventive risk management. Where practical, risk responses should be designed in a manner that complies with local regulation (e.g., the US Health Insurance Portability and Accountability Act [HIPAA]), international standards (e.g., *ISO/IEC 27001—Information Security Management*), industry standards (e.g., Payment Card Industry Data Security Standard [PCI DSS]) and internal standards promulgated by the organization. Organizations that comply with standards may derive competitive advantage from certification of their compliance.

1.5.2 IT Risk in Relation to Other Business Functions

IT risk is a critical part of business, as reliance on IT has grown exponentially. As a general practice, unless an enterprise is willing to take on risk, it will not be able to realize the benefits and rewards associated with pursuing a course of action. However, taking on too much risk or exceeding the risk capacity may not only lead to the loss of investments, but also jeopardize the business viability and longevity resulting in its catastrophic failure. Senior management is responsible not only for establishing the risk appetite and tolerance criteria for the enterprise—clear statements of how much risk to take and what opportunities to forego, but also ensuring it is communicated.

While the risk practitioner is primarily concerned with IT risk, this is just a subset of business risk. This requires the risk practitioner to understand the risk culture of an enterprise and to use it to drive or inform the IT risk strategy. The business does not exist so that the enterprise can have an IT department; the IT department exists to help the business achieve its strategic mission and meet its goals and objectives.

When analyzing and assessing IT risk, the risk practitioner must not calculate risk solely from the perspective of IT, rather they need to ensure that both technical and nontechnical elements of risk have been properly considered. A failure of an IT system has an impact on the IT department, and it most likely will have a much greater impact on the business which relies and depends on that system more than IT alone.

In addition to the relationship between business risk and IT risk, the risk practitioner should also be familiar with risk-related business functions that establish the foundations of organizational resiliency. The various management functions and goals of business continuity, audit, information security, controls, projects and change management need to be understood and integrated to gain perspective and ensure alignment with what the organization is both concerned about and values.

Risk and Business Continuity

IT risk management is closely linked with business continuity. The business function is concerned with the preservation of critical business functions and the ability of the enterprise to survive an adverse event that may impact the ability of the enterprise to meet its mission and goals. Through risk management, the enterprise attempts to reduce all IT risk to an acceptable level. Although the controls and efforts of IT risk management may not prevent a failure, the risk practitioner works with the incident management and business continuity teams to identify possible threats and put in place the mechanism to detect, contain and recover from an adverse event if it should happen.

Where available (depending on the enterprise's maturity), an established business continuity capability can act as a starting point for risk management efforts; as business impact analyses have already identified key processes, business criticality and identified what the organization has determined as acceptable and by inverse unacceptable losses to the enterprise. If the business continuity plan (BCP) is inadequate or inaccurate, the organization may not meet its goals for recovery after an incident.

Risk and Audit

The audit function is an important part of corporate governance that provides management with assurance regarding the effectiveness of the control framework, IT risk management program and compliance. In a world of increasing legislation, government oversight and media scrutiny, organizations must diligently demonstrate an adequate control environment and proactive risk management practices. For that reason, IS audits should be conducted by objective, skilled and independent personnel able to review risk, identify vulnerabilities, document findings and provide recommendations on how to address audit issues.

An audit is a methodical and structured review that requires competence and knowledge in the subject matter of the area being audited. If the IS auditor is not familiar with the technology being used, the significance of operating conditions or the requirements of the enterprise, the audit may be inaccurate and provide limited value. IS audits must also be independent. Senior management is often involved in the creation of the IS audit plan, and in a situation where a particular manager is involved in inappropriate activity, they may restrict the ability of IS audit to perform their duties effectively. Even when there is no wrongdoing, the appearance of partiality may cast doubt on the results of an audit, creating less value than expected. The risk practitioner should review the relationship between the IS auditor and the area being audited to ensure that there is no conflict of interest.

Risk and Information Security

IT risk management drives the selection of controls and justifies their initial and continued operation. If the IT risk management activity is not conducted properly, information security controls are almost certain to be incorrectly designed, poorly implemented and improperly operated. Every control should be traceable back to a specific IT risk that the control is designed to mitigate, and the risk practitioner should be able to demonstrate the purpose of each control and explain the reasoning behind its selection.

Control Risk

A control is chosen to mitigate a risk, but if the control is not operating correctly then the control may not prevent a failure or compromise, in addition to creating a false sense of security (**figure 1.23**). The selection of the wrong control, the incorrect configuration of the control, the improper operation of the control, the failure to monitor and review the control or the inadequacy of the control to address new threats may introduce the risk of control failure.

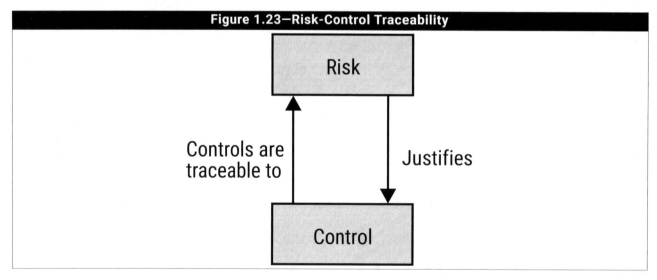

Figure 1.23—Risk-Control Traceability

When selecting a control, consider a number of factors for finding the most effective and efficient control. Understand that there are different types of controls and understand the capabilities and limitations of the controls being recommended.

Provide the risk owner with multiple control options, as providing a single option is effectively giving an ultimatum.

When recommending controls to mitigate risk, consider the following elements:

- Efficiency
- Effectiveness
- Exposure
- Implementation capability
- Organization

Project Risk

Many projects fail; in fact, numerous studies of IT projects have indicated that a majority of IT projects could be considered failures. Failure of a particular project may be defined by surpassing its allotted budget or the allotted time scheduled, or if it does not deliver what it promised. A project may also be deemed a failure if it delivered what it promised but the deliverables did not meet customer needs and expectations. The failure of an IT project may pose a significant risk to an enterprise, manifesting as lost market share, failure to seize new opportunities or other adverse impacts on customers, shareholders and staff. Identifying the risk associated with a project and successfully managing that risk is very likely to result in higher levels of project success and stakeholder satisfaction. See chapter 4 Information Technology and Security for more information.

Change Risk

Risk is not static. Changes in technology, regulations, business processes, functionality, architecture, users and other variables that affect the business and technical environments of the enterprise introduce a wide range of risk associated with systems in operation. The risk level of a particular system may also change because of intentional changes to its configuration or architecture that result in the controls that were originally effective as designed becoming ineffective.

The risk practitioner is tasked with managing risk on a continuous basis, which means that they must stay aware of emerging risk that may be associated with new threats, new technologies, changes in environment, culture and changes in legislation and/or regulation. All of these changes may affect the risk posture of the enterprise and result in a new level of risk not adequately addressed in earlier risk-identification efforts.

1.6 Organizational Assets

An asset is something of either tangible or intangible value that is worth protecting. Examples of assets commonly found in organizations include:

- Data, information and knowledge
- Reputation
- Brand
- Intellectual property
- Facilities
- Equipment
- Cash and investments
- Customer lists
- Research
- People
- Service/business process

CHAPTER 1— GOVERNANCE

1.6.1 People

Many enterprises are vulnerable to the loss of a key employee who may be the only person with knowledge in a certain area or specific expertise. Failures by management to identify key employees and ensure that they are supported through cross-training, sufficient documentation of key processes performed, and incentive programs are unfortunately common. Whether the loss occurs as a result of retirement, illness or recruitment by another organization, it may leave the enterprise in a precarious and vulnerable position.

1.6.2 Technology

Technology changes rapidly, and new technologies are always being developed. In addition to being aware of new technologies and the risk that they pose, the risk practitioner should also consider the risk associated with outdated technology, which is often overlooked. Equipment that is no longer supported or past its mean time between failure (MTBF) may be particularly vulnerable. Additionally, while contract vehicles such as "extended support" may extend system availability, they also result in higher operational costs over time. Lack of patching and updating of systems or applications leaves them vulnerable to malware or misuse, and older systems may require expertise that is not readily available to maintain (potentially increasing key-person dependency). Older systems may also lack sufficient documentation, may result in becoming reliant on one vendor to maintain or be difficult to support in terms of replacement parts.

When systems are scheduled for disposal, they may contain sensitive data. The risk practitioner should be sure that sufficient procedures are in place to securely dispose of such data and that these procedures are consistently followed. The method of secure disposal appropriate for a particular system is based in part by its physical form and on the sensitivity of the data on the device. Common methods of destroying data include overwriting, degaussing and physical destruction of the equipment. If the data may be needed later, the organization may need to retain a copy of the software or even a legacy system to read them. In the case of encrypted data, keys or password files must also be securely stored and managed. Failure to remove retired systems from backup schemes or business continuity or disaster recovery plans may affect those operations.

1.6.3 Data

Many enterprises consider data to be extremely valuable. Customer lists, financial data, marketing plans, human resources (HR) data and research are some examples of data-related assets that should be protected. The systems that host, process or transmit the data must ensure that data are protected at all times, in all forms (paper, magnetic storage, optical storage, reports, etc.) and in all locations (storage, networks, filing cabinets, archives, etc.).

To ensure the proper handling, use and safeguarding of data, the enterprise should clearly identify the business value of the data. The appropriate data security classification should also be defined, which describes the various data elements to be considered, the potential impact should the confidentiality, integrity or availability be compromised and the corresponding controls to ensure such risk is reduced to an acceptable level.

Figure 1.24 provides an example of a security categorization process.

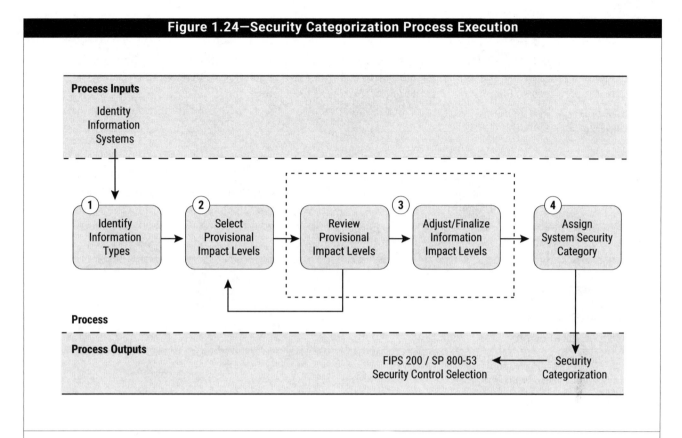

Figure 1.24—Security Categorization Process Execution

Source: National Institutes for Standards & Technology, *NIST SP 800-60: Volume I: Guide for Mapping Types of Information and Information Systems to Security Categories*, USA, 2008, figure 2

1.6.4 Intellectual Property

Trademarks, copyrights, patents, trade secrets and other items associated with the reputation and goodwill of the enterprise and research that leads to a new product, are examples of intellectual property. Intellectual property is a class of information that is treated with special care because it may represent the future earnings potential of the enterprise. Failure to protect intellectual property may result in the loss of competitive advantage. All employees and business partners should be bound by nondisclosure agreements (NDAs) and reminded of their responsibility to protect the intellectual property of the enterprise and handle it properly. This may include strict access controls, shredding of documents, caution when discussing information in a public location and encryption of data on portable media. Key terms related to intellectual property are described in **figure 1.25**.

Figure 1.25—Intellectual Property Terms	
Term	Definition
Trademark	A sound, color, logo, saying or other distinctive symbol that is closely associated with a certain product or company. Some trademarks are eligible for registration.
Copyright	Protection of any work that is captured in a tangible form (e.g., written works, recordings, images, software, music, sculpture, dance, etc.).
Patent	Protection of research and ideas that led to the development of a new, unique and useful product to prevent the unauthorized duplication of the patented item.
Trade secret	A formula, process, design, practice or other form of secret business information that provides a competitive advantage to the organization that possesses the information.

1.6.5 Asset Valuation

Not all assets are equally important; many are absolutely critical to business operations, while others provide indirect support and benefits. The risk practitioner should try to determine the importance of assets in the context of organizational activities so that priority may be given to protecting the most important assets first and addressing less significant assets as time and budget allow. Effective valuation also protects the organization from paying more in protection than the value of the asset.

The calculation of the value an asset provides to an enterprise is not as straightforward as it may initially appear. Many enterprises use a quantitative approach that assigns a monetary value, which can be difficult when the value influences or drives intangibles such as confidence, morale or market perception. For example, if a product, company or region is associated with poor quality, environmental negligence or fraudulent activity, a product associated with any of these factors may result in the product being perceived as substandard due to this market perception, regardless of the actual product. The stigma of negative perception may persist for many years before an enterprise can recover.

Contributing factors to calculating asset value include:

- Financial penalties for legal noncompliance
- Impact on business processes
- Damage to reputation
- Additional costs for repair/replacement
- Effect on third parties and business partners
- Injury to staff or other personnel
- Violations of privacy
- Breach of contracts
- Loss of competitive advantage
- Legal costs

When calculating asset value, one technique is to base it on the impact of a loss of confidentiality, integrity and availability (CIA). This approach attempts to relate the impact to easily understood terminology. In order for this to work, the values for each level must be clearly stated and used in the same way by all departments.

Asset Inventory and Documentation

An important part of managing risk is the identification and inventory of the assets of the organization. As changes are made to systems and new equipment is installed or older equipment is retired, the asset inventory must be updated to reflect what assets, systems and equipment the organization is currently using. For hardware asset inventories, they should list all equipment, supplier, acquisition date, original cost, actual cost, the location, owner of the equipment and other data required for maintenance, insurance and warranty purposes.

Data/information asset inventories should list the system(s), source, acquisition method, business use, business criticality, availability, completeness, processing, storage, transmission, sensitivity, classification and the business owner. Often the task of identifying, inventorying and mapping data can appear to be a herculean endeavor; however, it is a common requirement in a number of regulatory, statutory and contractual agreements and with the advent of legislation relating to privacy, such as the European Union's General Data Protection Regulation (EU GDPR), California Consumer Privacy Act (CCPA) and the PCI DSS. Additionally, it is referenced frequently as a requirement in industry recognized standards, such as ISO/IEC 27001, and common industry accepted best practice such as the NIST Cybersecurity Framework and Center for Internet Security's Top 20 Critical Security Controls.

Part B: Risk Governance

Risk management functions—the processes, practices and activities which fall within management's purview—need to be sufficiently governed to ensure continued alignment with the enterprise's goals and objectives. Where the governing body sets the enterprise strategy and direction relating to risk management capabilities, it falls to management to successfully deliver on that capability.

Governing risk requires alignment of risk management capabilities to support and deliver on the organization's expectations.

1.7 Enterprise Risk Management and Risk Management Frameworks

Risk management is defined as the coordinated activities to direct and control an enterprise with regard to risk. In simple terms, risk can be viewed as a challenge to achieving objectives, and risk management as the activity undertaken to predict challenges and lower their chances of occurring and/or their impact. Effective risk management can also assist in maximizing opportunities, and the risk practitioner should keep this upside/downside duality of risk in mind. For example, a risk decision might take the form of potential benefits that may accrue if opportunities are taken versus missed benefits if those same opportunities are foregone.

1.7.1 IT Risk Management Good Practices

The process of IT risk management is most reliably effective when it follows a structured methodology based on good practices and a desire to seek continuous improvement. The risk practitioner should begin a new risk management effort by reviewing current practices of the enterprise in the identification, assessment, response, monitoring and reporting of risk. On the basis of this initial evaluation, the risk practitioner can gain valuable insight into how the enterprise views risk management and identify areas in which the current program may incorporate or deviate from recognized good practices, which can facilitate the development of a consistent program.

Where good practices are not already in place, the risk practitioner may find it beneficial to either formally adopt or informally draw upon one or more well-established standards or frameworks, which can help to ensure that the risk management program is complete and authoritative.

The IT risk management program should be:

- Comprehensive (thorough, detailed)
- Complete (carried through to the end)
- Auditable (reviewable by an independent third party)
- Justifiable (based on sound reasoning)
- Compliant (with policy, laws and/or regulations)
- Monitored (subject to review and accountability)
- Enforced (consistent, mandated and required)
- Up to date (current with changing business processes, technologies and laws)
- Managed (adequately resourced, with oversight and support)

Good practices in IT risk management, like all other good practices, are subject to change and improvement. The risk practitioner should be aware of these changes and take steps to ensure that the organization adopts practices that are informed by the best available industry knowledge and experience, while keeping in mind that any standard or framework may need to be tailored to support the particular goals of the enterprise.

1.7.2 Establishing an Enterprise Approach to Risk Management

Risk management is an enterprise activity that benefits from a standardized, structured approach that can be applied to the entire enterprise without substantial modification or customization. It is possible to identify risk on a system-by-system or project-by-project basis, but the result of such an approach actually creates new risk of false assurance by having neither consistency nor interoperability among the risk solutions that are implemented. Without a structured approach, risk may be measured differently in different areas, creating gaps between boundaries of the various projects or systems.

This is not to say that risk should never be tailored. In a large enterprise consisting of many divisions, departments, lines of business and products or services, different cultures and business models may warrant tailoring to the needs of each organization. There is no one risk management approach that is suitable for all types of organizations or even for all components of a large, diverse organization. The risk practitioner must be sensitive to local departmental cultures, priorities, regulations, goals and restraints before recommending a risk management approach or framework. However, the recommended framework should remain substantially consistent across the organization even when it is tailored. In particular, it is important that the results of risk management in one division be comparable to the results in another.

In some large enterprises and governments, the risk management function serves a consultative and advisory role that provides data and recommendations to each department but does not actually determine the development of the control or risk mitigation framework. In such an enterprise, risk management is a clearinghouse for information that can be accessed by all departments as a part of systems or project development. This approach is legitimate and can be effective when risk is generally integrated throughout the culture of the enterprise.

Executive Sponsorship (Tone at the Top)

Enterprise risk management benefits strongly from the clear support from senior management, which should require consultation with risk practitioners to be part of any new project and ensure that recommendations of the risk management program are evaluated and objectively addressed before approving or funding projects or business initiatives.

Policy

A critical part of establishing the risk-management process is the development and approval of a concise, coherent risk management policy that reflects the attitude and intent of management in relation to risk. A risk management policy should include a statement relating to the reasoning or rationale behind the approach to accepting or mitigating risk, set accountability, and articulate a commitment to continuous improvement of the risk environment.

1.8 Three Lines of Defense

The three lines of defense has been implemented by a variety of organizations in order to better enhance enterprise risk management capabilities across the entirety of the organization's various lines of business, establishing a more robust ERM program.

This model identifies several business functions between *risk owner*—the primary individual who are ultimately accountable and responsible for how risk are addressed and *risk practitioner*—the individual who is delivering risk management functions. The risk practitioner is responsible for establishing core processes, practices, activities and roles relating to risk management capabilities. This includes clear delineation between governance and management responsibilities and providing for the appropriate level of risk *oversight* and *assurance* levels.

1.8.1 The First Line of Defense: Operational Management

The first line of defense is often implemented by the business unit, component or business function that performs daily operation activities, especially those that are the front lines of the enterprise. In this case they are expected to:

- Ensure the conductive control environment in their business unit.
- Implement risk management policies regarding roles and responsibilities, especially in activities that lead to corporate growth. They are expected to be fully aware of the risk factors that should be considered in every decision and action.
- Be able to execute effective internal control in their business units, as well as the monitoring process and maintaining transparency in the internal control itself.

As the first line of defense, operational managers (business owners) are responsible for managing risk. In short, they are the risk owners. They are responsible for implementing corrective actions to address risk that exceed the defined risk appetite. This includes the responsibility for controls that have been implemented to keep risk within tolerable levels.

1.8.2 The Second Line of Defense: Risk and Compliance Functions

The second line of defense is typically comprised of risk management and compliance functions, where the expectations are to:

- Establish business-aligned risk management frameworks.
- Develop, monitor and oversee the process and operation of the enterprise's overall risk management program capabilities.
- Monitor and ensure that business functions operate in accordance with the enterprise's risk management policies, standards, and procedures.
- Monitor and report to the appropriate stakeholders on the enterprise's current risk profile, posture and exposure, changes thereof, the status of any ongoing risk mitigation activities, and significant, credible threats faced by the enterprise.

Within the second line of defense, risk management functions are focused to ensure that first-line defenses are properly designed, implemented and operating as intended. While these functions have some degree of autonomy from the first line of defense, they are still management functions and introduce a level of subjectivity and bias as they may directly interact with the modification or development of controls and risk systems.

Additionally, one of the more important second line activities the risk practitioner should carry out is gaining enterprisewide consensus, buy-in and adoption of the risk management standard framework, as this will be used to by the third line of defense (audit) to assess effectiveness and efficiently. The establishment of an accepted risk management standard relating to the processes, practices and activities defines and provides the standard against which the risk management program will be assessed.

This is beneficial for several reasons. First, it reduces the level of effort required by audit in not only performing the audit, but also does not require audit to create their own standard to assess conformance. Second, it reduces potential false positives/findings associated with creating custom standards. Third, it focuses audit efforts based upon the enterprise's approved standard, not requiring attempting to apply industry good practices that may not align with the enterprise's goals and objective. Finally, it frames risk management functions as a business discussion, in that an audit finding which exists based on a perceived industry best practice allows the business to make the decision on whether or not to adopt said standard.

CHAPTER 1— GOVERNANCE

1.8.3 The Third Line of Defense: Audit

The third line of defense consists of gaining the appropriate level of assurances to senior management and the governing body through independent and objective reviews performed by the enterprise's internal audit capability and external audit (if applicable). An internal auditor is expected to:

- Assess conformance of the risk management program against the organizationally adopted and approved standard
- Review and evaluation the design and implementation of risk management holistically.
- Ensure the efficacy of the first and second lines of defense.

The risk practitioner may be called to speak to the design, implementation, effectiveness and efficiency of first- or second-line processes, practices, activities and controls and provide the appropriate context, based on the desired level of assurance required.

1.8.4 The Role of the Risk Practitioner within the Three Lines of Defense

The risk practitioner often works across the first two lines of defense within an organization, while audit provides the governing board and senior management the independence and objectivity needed to report on the conformance with the applicable standards and frameworks identified by the risk practitioner and adopted by the organization. **Figure 1.26** describes the three lines of defense.

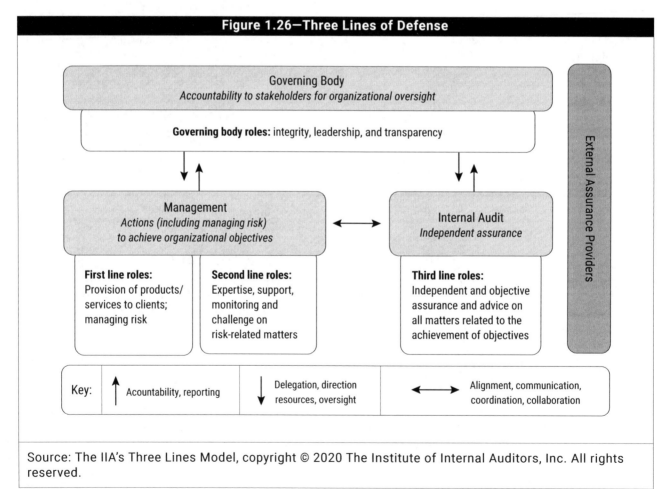

Source: The IIA's Three Lines Model, copyright © 2020 The Institute of Internal Auditors, Inc. All rights reserved.

Risk practitioners are uniquely positioned within the enterprise to aid in providing enterprise assurances to senior management, audit committee and boards of directors relating to risk governance and risk-management processes. By virtue of their role, the risk practitioner is positioned to act in an advisory capacity on the variety of threats and risk being faced, the analysis and assessment of risk, the selection of appropriate risk response and reporting on changes to the enterprise's risk posture, profile and overall exposure.

1.9 Risk Profile

Risk management functions exist to serve the needs of the enterprise. Because the enterprise evolves over time, risk management is an ongoing, cyclical process that recognizes the dynamic nature of risk and the need for continuous monitoring and assessment. Temporary stability may be attained as a result of program maturity, but the risk practitioner cannot become complacent. Changes in the organizational risk environment may prompt the complete reevaluation at any time.

The risk profile is based on the overall risk posture of the enterprise reflected in its attentiveness to monitoring the effectiveness of controls, proactive identification and addressing or preventing risk and development of a risk culture.

Changes in the risk profile may be the result of numerous factors, including:

- New technologies
- Changes to business procedures
- Mergers or acquisitions
- New or revised regulations
- Changes in customer expectations
- Actions of competitors
- Effectiveness of risk-awareness programs

Management should pursue the development and maturity of risk-management processes and culture with careful attention. Risk at the organization level is the product of risk as it exists at the levels of individual systems, facilities and process. Similar to health, safety and information security, the risk posture of an organization may be affected by the cascading effects of seemingly minor actions.

> **On the Topic: Mergers and Acquisitions**
>
> When an enterprise goes through a merger or acquisition, new risk will emerge. Mergers create uncertainty and stress that can result in poor judgment or inappropriate actions by people worried about protecting their jobs, being moved into other positions or having to compete for a new assignment. In many cases, mergers result in job loss, and some studies have suggested that employees who are about to lose their jobs are likely to take corporate assets and information with them. When employees have assets of the enterprise in their possession, including access to sensitive or protected information, the risk practitioner should actively increase the controls over sensitive information, be increasingly diligent to enforce monitoring and work with HR to identify potentially risk-laden scenarios.

To evaluate the risk control effectiveness and efficiency and determine the IT risk profile of the enterprise, risk practitioners should continually monitor an assortment of indicators including:

- New assets
- Changes to the scope of risk assessment
- Changes in business priorities, assets, services, products and operations

CHAPTER 1— GOVERNANCE

- Risk acceptance levels
- New threats (internal and external)
- Newly discovered vulnerabilities
- Possibility of increased risk due to aggregation of threats and vulnerabilities
- Incidents
- Logs and other data sources
- People and the morale of the organization
- Regulations and legal changes

Other areas that may be candidates for monitoring include:

- Actions of competitors (benchmarking and standards)
- Changes in the supply chain (e.g., mergers, political turmoil, transportation problems, natural events, bankruptcies of vendors)
- Changes in the financial markets
- Total cost of ownership (TCO) of assets
- Impact from external events
- Availability of staff/resources

IT risk management objectives and goals should be reviewed to ensure that they continue to be aligned with the goals and objectives of senior management on a regular basis. Many organizations find it beneficial to perform the review on an annual basis and to include criteria for monitoring, thresholds used for KPIs and KRIs, policies and strategies of risk, the reporting schedule and the list of key stakeholders to be notified when KPIs or KRIs exceed their thresholds. The review also provides a valuable opportunity to review the program in terms of increasing maturity, including the completion of risk response and mitigation activities, the training of staff, the success of awareness programs, improved response time to incidents, timely rollout of patches, and better alignment and communication among management, audit, business continuity, physical security and information security departments.

Risk is owned by management, but the risk practitioner has a key role in ensuring that management is aware of the current IT risk profile and that risk is being managed in a way that meets management objectives. Throughout this phase of the IT risk-management process, the risk practitioner works with risk owners, IT staff, third parties, incident response teams and auditors to monitor risk and evaluate the effectiveness and efficiency of the control framework. As incidents occur, lessons learned are used to improve the risk-management process through better knowledge, staffing, technical controls, procedures, monitoring and response programs. These benefits can help avoid future problems, minimize the impact of future incidents and sustain business operations.

1.10 Risk Appetite, Tolerance and Capacity

Every enterprise has a particular risk capacity, defined as the objective amount of loss an enterprise can accept without its continued existence being called into question. Subject to the absolute maximum imposed by this risk capacity, the owners or board of directors set the risk appetite for the enterprise. Risk appetite is defined as the amount of risk, on a broad level, that an entity is willing to accept in pursuit of its mission. In some cases, setting the risk appetite may be delegated by the board of directors to senior management as part of strategic planning. The relationship between risk appetite, risk tolerance and actual risk are shown in **figure 1.27**.

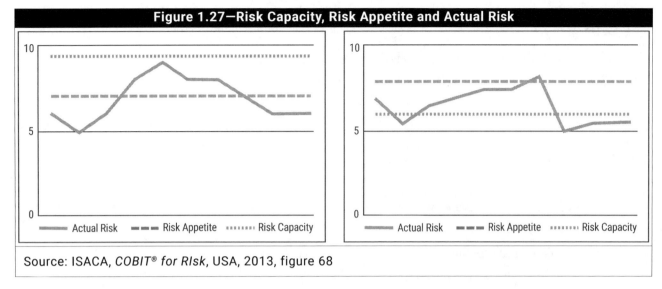

Figure 1.27—Risk Capacity, Risk Appetite and Actual Risk

Source: ISACA, *COBIT® for RIsk*, USA, 2013, figure 68

Some benefits of defining risk capacity and risk appetite at the enterprise level include:

- Supporting and providing evidence of the risk-based decision-making processes
- Supporting the understanding of how each component of the enterprise contributes to the overall risk profile
- Showing how different resource allocation strategies can add to or lessen the burden of risk by simulating different risk response options
- Supporting the prioritization and approval process of risk response actions through risk budgets
- Identifying specific areas where a risk response should be made

Risk appetite is translated into a number of standards and policies to contain the risk level within the boundaries set by the risk appetite. These boundaries need to be regularly adjusted or confirmed. Within these boundaries, risk may be accepted through a formal and explicit process that affirms that the risk requires and warrants no additional response by the enterprise as long as it and the risk environment stay substantially the same and accountability for the risk is assigned to a specific owner.

Risk acceptance generally should not exceed the risk appetite of the enterprise, but it must not exceed the risk capacity (which would threaten the continued existence of the enterprise).

However, there are circumstances where deviations above the defined risk appetite are permissible, as long as they meet specific criteria, these are known as the risk tolerance. These deviations are not desirable but are known to be sufficiently below the risk capacity; acceptance of risk is still possible when there is compelling business need and other options are too costly. Risk tolerance may be defined using IT process metrics or adherence to defined IT procedures and policies, which are a translation of the IT goals that need to be achieved. Like risk appetite, risk tolerance is defined at the enterprise level and reflected in the policies created by senior management. Exceptions can be tolerated at lower levels of the enterprise as long as the overall exposure does not exceed the risk appetite at the enterprise level.

Note: Risk tolerance is defined as the acceptable level of variation that management is willing to allow for any particular risk as the enterprise pursues its objectives. The interpretation of the ISACA definition is that while management has an official acceptance level of one value, they may accept a slight deviation from that level. An example of tolerance is a situation where the speed limit on a highway is 65 miles/hour (105 kilometers/hour), but a police officer may allow a person to travel up to 70 miles/hour (113 kilometers/hour) before issuing a ticket.

Risk appetite and tolerance need to be defined and approved by senior management and clearly communicated to all stakeholders, with a process in place to review and approve any exceptions at the appropriate level of authority.

As with all risk, risk appetites and tolerances change over time, and a variety of factors such as new technology, organizational restructuring, or changes in business strategy may require the enterprise to reassess its risk portfolio and reconfirm its risk appetite.

Creating a risk-aware culture is an important aspect of monitoring risk. This may be a charge of senior management through revisiting and reinforcing the risk appetite and holding the proper stakeholders accountable for implementing risk management that align with the enterprise's risk appetite.

According to the Committee of Sponsoring Organizations of the Treadway Commission (COSO), this is best accomplished when the enterprise communicates the risk appetite and risk tolerance, with the following outcomes:[7]

- Consistent implementation across units
- Effective monitoring and communication of risk and changes in risk appetite
- Consistent understanding of risk appetite and related tolerances for each organizational unit
- Consistency between risk appetite, objectives and relevant reward systems

1.11 Legal, Regulatory and Contractual Requirements

Enterprises are required to comply with the laws and regulations of the jurisdictions in which they operate and face penalties for failing to do so. It is important to know what laws apply to the enterprise and to understand their requirements, which can be challenging because many laws are open to interpretation and required levels of compliance are not always stipulated. For example, a law may require adequate protection of sensitive data without specifying what constitutes an adequate level of protection. Laws and regulations may also contradict one another, and an enterprise that operates in various regions of the world may find it difficult to comply with all laws in each location.

Enterprises that operate globally or even within different regions of one country may build a global program of policies and a control suite to handle the common regulations, and then have a regional or nation-specific addendum to handle the exceptions and their controls.

For instance, the European Union General Data Protection Regulation states:[8]

Personal data must be:

1. *Processed lawfully and fairly;*
2. *Collected for specified, explicit and legitimate purposes and not processed in a manner that is incompatible with those purposes;*
3. *Adequate, relevant and not excessive in relation to the purposes for which they are processed;*
4. *Accurate and, where necessary, kept up to date; every reasonable step must be taken to ensure that personal data that are inaccurate, having regard to the purposes for which they are processed, are erased or rectified without delay;*
5. *Kept in a form which permits identification of data subjects for no longer than is necessary for the purposes for which they are processed;*
6. *Processed in a manner that ensures appropriate security of the personal data, including protection against unauthorized or unlawful processing and against accidental loss, destruction or damage, using appropriate technical or organizational measures.*

CHAPTER 1 — GOVERNANCE

Regulations may require enterprises to report on their own compliance and impose financial penalties or loss of a license to operate if these reports are made incorrectly or outside of a directed schedule. The risk practitioner who is tasked with either building such a program or administering one already in place should work with legal counsel and other professionals across the organization to ensure an adequate understanding of what is expected. Reports on compliance should be accurate, complete and submitted in a timely manner. To ensure compliance, an organization must have the ability to monitor and measure the controls in use. Reports should be comparable from one reporting period to another, and any trends or areas of noncompliance should be identified and addressed. In cases where there is noncompliance, a justification of the reasons for noncompliance should be documented and provided on request.

In some cases, compliance may also apply to voluntary standards. For instance, the PCI DSS, created by members of the payment card industry, is not required by law but is generally required by the issuers of payment cards to be adopted by companies that want to handle payment cards. **Figure 1.28** list the twelve PCI DSS requirements as an example of the types of requirements that a standard may impose on an enterprise.

Figure 1.28—PCI Data Security Standard—High Level Overview	
The 12 PCI DSS Requirements	
Build and maintain a secure network and systems	1. Install and maintain a firewall configuration to protect cardholder data.
	2. Do not use vendor-supplied defaults for system passwords and other security parameters.
Protect cardholder data	3. Protect stored cardholder data.
	4. Encrypt transmission of cardholder data across open, public networks.
Maintain a vulnerability management program	5. Protect all systems against malware and regularly update anti-virus software or programs.
	6. Develop and maintain secure systems and applications.
Implement strong access control measures	7. Restrict access to cardholder data by business need to know.
	8. Identify and authenticate access to system components.
	9. Restrict physical access to cardholder data.
Regularly monitor and test networks	10. Track and monitor all access to network resources and cardholder data.
	11. Regularly test security systems and processes.
Maintain an information security policy	12. Maintain a policy that addresses information security for all personnel.
Source: Provided courtesy of PCI Security Standards Council, LLC and/or its licensors. ©2006-2021 PCI Security Standards Council, LLC. All Rights Reserved. *Payment Card Industry (PCI) Data Security Standard*, v3.2.1	

The risk practitioner should always keep in mind that compliance is a risk decision, even when it is imposed by force of law. As discussed earlier, risk is addressed in the most cost-effective manner possible based on the risk appetite set by senior management. Depending on the penalties attached to noncompliance, an organization may choose not to be compliant with certain laws or regulations if the cost of compliance is greater than the fine or consequences imposed for failure to comply.

1.12 Professional Ethics of Risk Management

Risk is often impacted by professional ethics. It is easy to understand that an organization with poor ethical standards may be more susceptible to fraud or theft, but ethical risk may also apply to an enterprise that has poor management

CHAPTER 1— GOVERNANCE

processes in place to identify errors, misuse or fraud. Ethics are related to an individual's perception of right and wrong and are not necessarily linked to the law. For example, in many organizations, it is acceptable to receive gifts from clients or suppliers, whereas in others it is not acceptable. To address the risk of a person violating the ethics policy of the enterprise, senior management must communicate policy to everyone and ensure that it is both visibly enforced and equally applicable to people at all levels of authority. Ethics policies that are unenforced or applied selectively are ineffective in directing behavior.

Ethics also applies to how people believe that they have been treated. A well-treated employee may be an active supporter of proper behavior. In contrast, an employee who feels that he or she has been poorly treated may seek revenge, with serious consequences.

Additionally, certain industries have incorporated ethics into their expectations, which can then result in both reporting and conformance requirements being established for various professionals.

As a risk practitioner who has earned the CRISC certification, ISACA has defined and set forth a Code of Professional Ethics to the professional conduct of members of the association and/or its certification holders.

ISACA certification holders shall:[9]

1. Support the implementation of, and encourage compliance with, appropriate standards and procedures for the effective governance and management of enterprise information systems and technology, including audit, control, security and risk management.
2. Perform their duties with objectivity, due diligence and professional care, in accordance with professional standards.
3. Serve in the interest of stakeholders in a lawful manner, while maintaining high standards of conduct and character, and not discrediting their profession or the Association.
4. Maintain the privacy and confidentiality of information obtained in the course of their activities unless disclosure is required by legal authority. Such information shall not be used for personal benefit or released to inappropriate parties.
5. Maintain competency in their respective fields and agree to undertake only those activities they can reasonably expect to complete with the necessary skills, knowledge and competence.
6. Inform appropriate parties of the results of work performed including the disclosure of all significant facts known to them that, if not disclosed, may distort the reporting of the results.
7. Support the professional education of stakeholders in enhancing their understanding of the governance and management of enterprise information systems and technology, including audit, control, security and risk management.

---

[1] ISO; *ISO 31000:2018 Risk Management—Principles and Guidelines*, Switzerland, 2018
[2] ISO; *IEC 31010:2019 Risk Management—Risk Assessment Techniques*, Switzerland, 2009
[3] ISO; *ISO/IEC 27001:2013 Information Technology—Security Techniques—Information Security Management Systems—Requirements*, Switzerland, 2013
[4] International Organization for Standardization/International Electrotechnical Commission (ISO/IEC); *ISO/IEC 27005:2018 Information Technology—Security Techniques—Information Security Risk Management*, Switzerland, 2011.
[5] National Institute of Standards and Technology (NIST); *NIST Special Publication 800-30 Revision 1: Guide for Conducting Risk Assessments*, USA, 2012
[6] NIST; *NIST Special Publication 800-39: Managing Information Security Risk*, USA, 2011
[7] Rittenburg, Larry; Frank Martens; *Enterprise Risk Management—Understanding and Communicating Risk Appetite*, Committee of Sponsoring Organizations of the Treadway Commission (COSO), USA, 2012
[8] Council of the European Union, *Global Data Protection Regulation*, 2016, https://eur-lex.europa.eu/legal-content/EN/TXT/HTML/?uri=CELEX:32016L0680&from=EN#d1e1178-89-1
[9] ISACA, "Code of Professional Ethics," https://www.isaca.org/credentialing/code-of-professional-ethics

Chapter 2:
IT Risk Assessment

Overview

Domain 2 Exam Content Outline 74
Learning Objectives/Task Statements 74
Suggested Resources for Further Study 75

Part A: IT Risk Identification

2.1 Risk Events 80
2.2 Threat Modeling and Threat Landscape 86
2.3 Vulnerability and Control Deficiency Analysis 95
2.4 Risk Scenario Development 106

Part B: IT Risk Analysis, Evaluation and Assessment

2.5 Risk Assessment Concepts, Standards and Frameworks 117
2.6 Risk Register 123
2.7 Risk Analysis Methodologies 125
2.8 Business Impact Analysis 127
2.9 Inherent, Residual and Current Risk 131

CHAPTER 2— IT RISK ASSESSMENT

Overview

Risk practitioners analyze and evaluate IT risk to determine the likelihood and impact on business objectives to enable risk-based decision making. This chapter discusses IT risk identification, including understanding the organization's threat landscape and emerging risk, and then how to analyze and evaluate that risk once it has been identified.

This chapter represents 20 percent (approximately 30 questions) of the CRISC exam.

Domain 2 Exam Content Outline

A. IT Risk Identification

1. Risk Events
2. Threat Modeling and Threat Landscape
3. Vulnerability and Control Deficiency Analysis
4. Risk Scenario Development

B. IT Risk Analysis and Evaluation

1. Risk Assessment Concepts, Standards and Frameworks
2. Risk Register
3. Risk Analysis Methodologies
4. Business Impact Analysis
5. Inherent and Residual Risk

Learning Objectives/Task Statements

Upon completion of this chapter, the risk practitioner will be able to:

1. Identify potential or realized impacts of IT risk to the organization's business objectives and operations.
2. Identify threats and vulnerabilities to the organization's people, processes and technology.
3. Evaluate threats, vulnerabilities and risk to identify IT risk scenarios.
4. Establish and maintain the IT risk register and incorporate it into the enterprise-wide risk profile.
5. Facilitate the identification of risk appetite and risk tolerance by key stakeholders.
6. Promote a risk-aware culture by contributing to the development and implementation of security awareness training.
7. Conduct a risk assessment by analyzing IT risk scenarios and determining their likelihood and impact.
8. Identify the current state of existing controls and evaluate their effectiveness for IT risk mitigation.
9. Review the results of risk analysis and control analysis to assess any gaps between current and desired states of the IT risk environment.
10. Collaborate with control owners on the selection, design, implementation and maintenance of controls.
11. Collaborate with control owners on the identification of key performance indicators (KPIs) and key control indicators (KCIs).
12. Review the results of control assessments to determine the effectiveness and maturity of the control environment.

13. Conduct aggregation, analysis and validation of risk and control data.
14. Report relevant risk and control information to applicable stakeholders to facilitate risk-based decision-making.
15. Evaluate emerging technologies and changes to the environment for threats, vulnerabilities and opportunities.
16. Evaluate alignment of business practices with risk management and information security frameworks and standards.

Suggested Resources for Further Study

Chapman, Robert J.; *Simple Tools and Techniques for Enterprise Risk Management, Second Edition*, John Wiley & Sons Inc., USA, 2012

Committee of Sponsoring Organizations of the Treadway Commission; *Enterprise Risk Management: Understanding and Communicating Risk Appetite,* USA, 2012

Committee of Sponsoring Organizations of the Treadway Commission; *Enterprise Risk Management for Cloud Computing,* USA, 2012

Committee of Sponsoring Organizations of the Treadway Commission; *Enterprise Risk Management: Integrating with Strategy and Performance – Executive Summary,* USA, 2017

Duke, Annie; *Thinking in Bets: Making Smarter Decisions When You Don't Have All the Facts,* Random House, USA 2018

Freund, Jack; Jones, Jack; *Measuring and Managing Information Risk*, Butterworth-Heinemann, USA, 2014

Jordan, Ernest; Luke Silcock; *Beating IT Risks*, John Wiley & Sons Inc., USA, 2005

Hubbard, Douglas; *How to Measure Anything 3rd edition*, Wiley, USA, 2014

Hubbard, Douglas; *How to Measure Anything in Cybersecurity Risk*, Wiley, USA 2016

Hubbard, Douglas; *The Failure of Risk Management*, Wiley, USA, 2009

ISACA, *COBIT 2019: Introduction and Methodology,* USA, 2018, www.isaca.org/cobit

ISACA, *COBIT Focus Area: Information Risk*, USA, 2021

ISACA, *Cybersecurity Fundamentals Study Guide*, 3rd edition, USA, 2021

ISACA, *The Risk IT Framework 2nd Edition*, USA, 2020

ISACA, T*he Risk IT Practitioner Guide 2nd Edition*, USA, 2020

National Institute of Standards and Technology, *NIST Special Publication 800-30 Revision 1: Guide to Conducting Risk Assessments,* USA, 2012

National Institute of Standards and Technology, *NIST Special Publication 800-39: Managing Information Security Risk*, USA, 2011

Taleb, Nassim Nicholas; *The Black Swan Second Edition: The Impact of the Highly Improbable*, Random House, USA, 2010

CHAPTER 2— IT RISK ASSESSMENT

Westerman, George; Richard Hunter; *IT Risk: Turning Business Threats Into Competitive Advantage*, Harvard Business School Press, USA, 2007

SELF-ASSESSMENT QUESTIONS

CRISC self-assessment questions support the content in this manual and provide an understanding of the type and structure of questions that have typically appeared on the exam. Questions are written in a multiple-choice format and designed for one best answer. Each question has a stem (question) and four options (answer choices). The stem may be written in the form of a question or an incomplete statement. In some instances, a scenario or a description problem may also be included. These questions normally include a description of a situation and require the candidate to answer two or more questions based on the information provided. Many times, a question will require the candidate to choose the **MOST** likely or **BEST** answer among the options provided.

In each case, the candidate must read the question carefully, eliminate known incorrect answers and then make the best choice possible. Knowing the format in which questions are asked, and how to study and gain knowledge of what will be tested, will help the candidate correctly answer the questions.

1. Which of the following statements **BEST** describes the value of a risk register?

    A. It captures the risk inventory.
    B. It drives the risk response plan.
    C. It is a risk reporting tool.
    D. It lists internal risk and external risk.

2. The **MOST** significant drawback of using quantitative risk analysis instead of qualitative risk analysis is the:

    A. lower objectivity.
    B. greater reliance on expertise.
    C. less management buy-in.
    D. higher cost.

3. Risk scenarios are analyzed to determine the:

    A. strength of controls.
    B. likelihood and impact.
    C. current risk profile.
    D. scenario root cause.

4. The **PRIMARY** reason risk assessments should be repeated at regular intervals is:

    A. omissions in earlier assessments can be addressed.
    B. periodic assessments allow various methodologies.
    C. business threats are constantly changing.
    D. they help raise risk awareness among staff.

5. Which of the following choices **BEST** helps identify information systems control deficiencies?

    A. Gap analysis
    B. The current IT risk profile
    C. The IT controls framework
    D. Countermeasure analysis

Answers on page 77

Chapter 2 Answer Key

Self-assessment Questions

1. A. A risk register is used to provide detailed information on each identified risk such as risk owner, details of the scenario and assumptions, affected stakeholders, causes/indicators, information on the detailed scores (i.e., risk ratings) on the risk analysis and detailed information on the risk response (e.g., action owner and the risk response status, time frame for action, related projects, and risk tolerance level). These components can also be defined as the risk universe.
   B. Risk registers serve as the main reference for all risk-related information, supporting risk-related decisions such as risk response activities and their prioritization.
   C. Risk register data are used to generate management reports but are not in themselves a risk reporting tool.
   D. The risk register tracks all internal and external risk, the quality and quantity of the controls and the likelihood and impact of the risk

2. A. Neither of the two risk analysis methods is fully objective. While the qualitative method subjectively assigns high, medium and low frequency and impact categories to a specific risk, subjectivity within the quantitative method is often expressed in mathematical weights.
   B. To be effective, both processes require personnel who have a good understanding of the business.
   C. Quantitative analysis generally has a better buy-in than qualitative analysis to the point where it can cause overreliance on the results.
   D. Quantitative risk analysis is generally more complex and, therefore, more costly than qualitative risk analysis.

3. A. The strength of controls is determined after the controls are in place to ensure they are adequate in addressing the risk.
   B. Risk scenarios are descriptions of events that can lead to a business impact and are evaluated to determine the likelihood and impact should the event occur.
   C. The current risk profile is the identification of risk currently of concern by the enterprise.
   D. The risk scenario process is used to identify plausible scenarios and from there determine likelihood and impact. Determining a root cause is not a part of the risk scenario process.

4. A. Performing risk assessments on a periodic basis can find omissions in earlier assessments, but this is not the primary reason for conducting regular reassessments.
   B. Enterprises strive to improve their risk-management process to more quickly and accurately assess and address risk, and this may involve changing the methodology. However, it is not the primary reason for conducting regular assessments.
   C. As business objectives and methods change, the nature and relevance of threats also change. This is the primary reason to conduct periodic risk assessments.
   D. Risk assessments are conducted on a periodic basis to address new threats and changes in the business. Creating more risk awareness is a minor benefit of conducting periodic risk assessments.

5. **A. Controls are deployed to achieve the desired control objectives based on risk assessments and business requirements. The gap between desired control objectives and actual IS control design and operational effectiveness identifies IS control deficiencies.**
   B. Without knowing the gap between desired state and current state, one cannot identify the control deficiencies.

CHAPTER 2—IT RISK ASSESSMENT

C. The IT controls framework is a generic document with no information such as desired state of IS controls and current state of the enterprise; therefore, it will not help in identifying IS control deficiencies.

D. Countermeasure analysis only helps in identifying deficiencies in countermeasures, not in the full set of primary controls.

Part A: IT Risk Identification

The risk identification process seeks to improve confidence that the enterprise recognizes and understands any risk with the potential to jeopardize its objectives. The process seeks to identify loss-event scenarios that may affect enterprise mission and strategic objectives. As shown in **Figure 2.1**, identification should leverage the monitoring and reporting that occur in the last process of the life cycle, which may alert the risk practitioner to changes in the risk environment and the risk practitioner's ability to identify emerging threats not previously known.

Source: Adapted from ISACA, *Getting Started with Risk Management*, USA, 2018, figure 2

Risk identification is important because only risk that is identified can be assessed and treated appropriately. When the enterprise fails to identify a risk, it is not included in the strategic planning process used by senior management to promote value creation. The risk practitioner should work closely with business process owners to ensure that they understand the ways in which the enterprise operates that is broad enough to include the enterprise and its external dependencies and assumptions, such as availability of contract labor or delivery of materials on a just-in-time basis. The risk practitioner should also consult with the various information technology and cybersecurity functions to understand the development, acquisition, implementation and integration processes for the wide range of technologies used by the enterprise and the purpose of these technologies. In addition, they should also understand how data and information are processed and handled along with the purpose of the data.

Given the combined reliance on and the intrusive nature of IT within most enterprises, this cannot be stressed enough, as most enterprises depend on IT in order to achieve their business objectives and goals.

CHAPTER 2— IT RISK ASSESSMENT

2.1 Risk Events

Risk events are those discrete, specific occurrences that result in an impact upon an enterprise. These differ from threat events as threats are the action events that could occur, and threat events describe the series of actions that may take place. However, on their own, they do not consider or include the likelihood of occurrence or impact to the enterprise. Threat events are often improperly categorized as "risk." However, in order to be a true risk event, the action (threat) must be combined with the event's potential likelihood and impact. Threats and risk come in various forms and originate from different actors; however, risk events consider the likelihood and impact on an enterprise's asset. Risk events can impact and act against specific targeted assets or intended to be broadly disruptive in execution.

Common examples of risk events can include:

- New regulations that impose new requirements on an enterprise
- Loss of key personnel
- Wildfires, hurricanes, flood or other natural disasters
- Fire at a data center
- Network intrusion by a cybercriminal
- Ransomware attack
- Abuse of positional authority

Risk identification requires the documentation and analysis of the elements that comprise risk such as:

- Consequences associated with specific assets
- Threats to those assets, normally requiring both intent (motivation) and capability
- Vulnerabilities that a threat may attempt to exploit
- Frequency that a potential risk event may occur (loss event) against an asset within the environment
- The potential harm that may result (loss event) against the asset if the threat were successful

Each element of risk needs to be considered both individually and in aggregate. Identification of risk depends on successful identification of assets, threats to those assets and the vulnerabilities that could be present that would allow an asset to be compromised. Key terms for the identification of threats and vulnerabilities are listed in **Figure 2.2**.

Figure 2.2—Risk Identification Terminology	
Term	Definition
Asset	Something of either tangible or intangible value that is worth protecting, including people, information, infrastructure, finances and reputation.
Asset value	The value of an asset to both the business and to competitors.
Impact	Magnitude of loss resulting from a threat exploiting a vulnerability.
Impact analysis	A study to prioritize the criticality of information resources for the enterprise based on costs (or consequences) of adverse events. In an impact analysis, threats to assets are identified and potential business losses determined for different time periods. This assessment is used to justify the extent of safeguards that are required and recovery time frames. This analysis is the basis for establishing the recovery strategy.
Impact assessment	A review of the possible consequences of a risk.
Likelihood	The probability of something happening.
Threat	Anything (e.g., object, substance, human) that is capable of acting against an asset in a manner that can result in harm.

Figure 2.2—Risk Identification Terminology (cont.)

Term	Definition
Threat actor	The entity which exploits a vulnerability, and includes various factors such as determination, capability, motive and resources.
Threat analysis	An evaluation of the type, scope and nature of events or actions that can result in adverse consequences; identification of the threats that exist against enterprise assets.
Threat vector	The path or route used by the adversary to gain access to the target.
Vulnerability	A weakness in the design, implementation, operation or internal control of a process that could expose the system to adverse threats from threat events.
Vulnerability analysis	A process of identifying and classifying vulnerabilities.
Vulnerability scanning	An automated process to proactively identify security weaknesses in a network or individual system.

2.1.1 Risk Factors

Risk is a combination of several factors that interact to cause damage to the assets of the enterprise, as shown in **Figure 2.3**.

CHAPTER 2— IT RISK ASSESSMENT

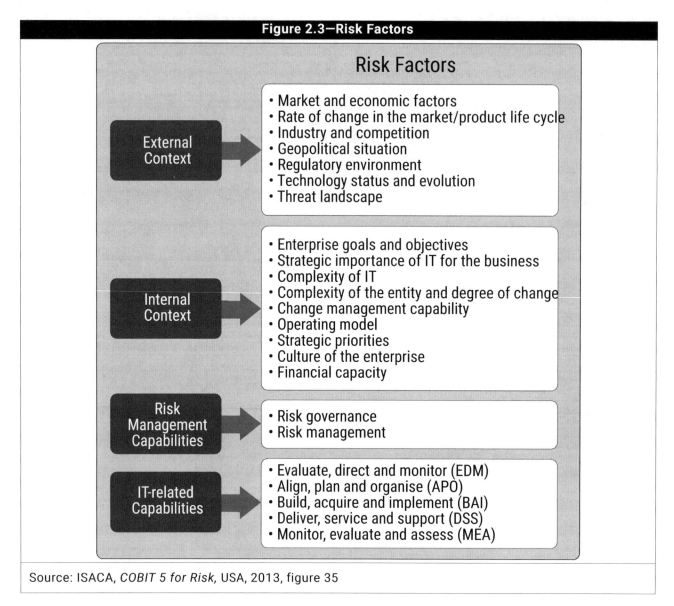

Source: ISACA, *COBIT 5 for Risk*, USA, 2013, figure 35

Risk factors influence the frequency and/or business or mission impact of risk scenarios; risk factors can be of different types and are classified into two major categories:

- **Contextual factors (internal or external)**—The main difference is the degree of control that an enterprise has over the respective factors.
   - Internal contextual factors are, to a large extent, under the control of the enterprise, although they may not always be easy to change.
   - External contextual factors are, to a large extent, outside of the control of the enterprise.
- **Capability factors (indicating ability to perform IT-related activities)**—These factors are critical to successful outcomes in managing risk. Capability factors are embedded in many related ISACA tools, techniques, methods and frameworks that support an enterprise in defining and improving IT and related processes needed to continue operating IT-related activities. Capability factors help answer these questions:
   - IT-related risk-management capabilities—To what extent is the enterprise mature in performing risk management?

- IT-related business or mission capabilities (or value management)—How robustly do IT-related capabilities support enterprise objectives while managing the risk that can jeopardize objectives?

As shown in **Figure 2.4**, a threat actor performs specific activities that result in an event occurring (an action), with the threat agent using one or many threat events in order to carry out their attacks against perceived vulnerabilities within an organization against assets that fall within their objectives. Threat actors that lack the ability to act may not pose a significant risk. Similarly, properly protected assets that are not vulnerable to threat events present no risk from attack. For example, if a virus has been written to exploit a vulnerability on a system, the virus can only have an impact if the system has not been patched to eliminate the vulnerability. After the patch has been applied, the virus poses no threat to the system.

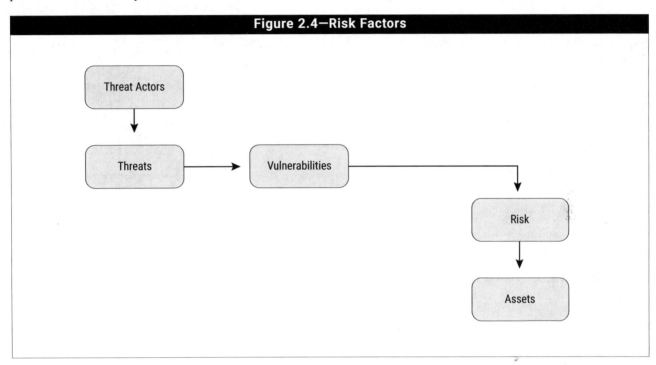

The first step in avoiding a threat is identifying it. Knowledge of threats and the motivations, strategy and techniques of those who perpetrate threats is key, so that a threat may be managed before it becomes reality. An enterprise needs to know its own weaknesses, strengths, vulnerabilities and the gaps in the security fabric. Security must be woven into each business process, each IT system and all operational procedures. Any gaps in the security fabric can easily be exploited. The better the risk practitioner understands the mind of the attacker or the source of the threat, the more effective the risk management activities will be in countering the threat.

The threat and vulnerability landscape is always changing. People move, equipment wears out, controls weaken, new threats emerge and security awareness dulls through inaction. The risk practitioner must proactively seek out threats, vulnerabilities and controls through regular assessments, testing, observation and analysis in order to be able to identify risk. Risk logs can be used as a tool to manage the risk generated during the identification process. A risk log can feature any kind of risk, and how risk logs are created and organized depends on the nature of the business the resources available. Because risk is cyclical and ongoing, risk logs are living documents, which need to be updated to ensure the entire enterprise understands how risk is being approached and managed. Establishing the right framework and right tools for managing risk logs is critical to ongoing risk management. A mapping of the trend of risk identified and risk in focus helps to serve as a metric to support key risk indicators.

Risk analysis is the modeling of various threats against assets, estimating the probability of a loss event occurring and the resulting impact on an asset, if the risk event were to be realized within an environment, and needs to consider technical and non-technical threats. When performing a risk analysis critical thought needs to be given relating to the various elements:

- The context, criticality and sensitivity of the system or process being reviewed
- The dependencies and requirements of the system or process being reviewed
- The operational procedures, configuration and management of the system or technology
- The training of the users and administrators
- The effectiveness of the controls and monitoring of the system or business process
- The manner in which data and system components are decommissioned

Risk evaluation is the consideration of risk events identified within the analysis, in context of the enterprise's defined risk appetite, tolerance criteria and capacity, and should consider the entire risk environment when evaluated. Evaluation of risk should include elements such as:

- Threat event frequency
- Loss event frequency
- Impact/loss magnitude
- Risk appetite
- Risk tolerance criteria
- Risk capacity

Risk is often influenced more by lack of training than by lack of equipment. In many cases, the risk is related to the way the equipment is operated more than whether the right tools are available.

2.1.2 Methods of Risk Identification

The risk practitioner has several possible sources for identification of risk, including:

- Historical- or evidence-based methods, or review of historical events, such as:
    - Audit or incident reports
    - Public media (e.g., newspapers, television, etc.)
    - Annual reports and press releases
- Systematic approaches (expert opinion) in which a risk team examines and questions a business process systematically to determine the potential points of failure, such as:
    - Vulnerability assessments
    - Review of business continuity and disaster recovery plans
    - Interviews and workshops with managers, employees, customers, suppliers and auditors
- Inductive methods (theoretical analysis), where a team examines a process to determine the possible point of attack or compromise, such as penetration testing
- Existing taxonomy, where a risk library with indicative IT-related risk can be used as a starting point. These libraries are commonly used in governance, risk and compliance (GRC) systems or risk assessment systems and support an integrated view of system, covering all aspects and risk sources.

The process of risk identification serves as the input to another aspect of risk management, as seen in **figure 2.5**.

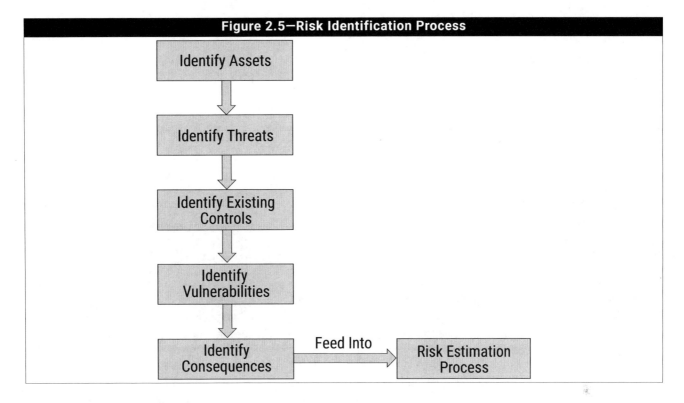

Figure 2.5—Risk Identification Process

2.1.3 Changes in the Risk Environment

The risk environment changes as organizations and technologies change and threat actors focus on new areas to attack. Systems and applications that were once thought to be secure are often found to be vulnerable at a later time. This requires the ongoing monitoring and the evaluation and assessment of the corresponding risk elements (see Chapter 3 Risk Response and Reporting).

Substantial changes in the risk environment may arise as a result of changes in technology or in business practices. The risk practitioner should ensure that new and emerging risk is identified, analyzed, evaluated, assessed and reported, and that the organization is aware of and watching for emerging threats and newly identified vulnerabilities. This includes monitoring vendor alerts, reports from computer emergency response teams (CERTs) and media stories. When a new issue arises, the risk practitioner should work with the business and system owner to perform a threat analysis and determine if and how the organization should respond.

Operational Integrity

Maintaining operational integrity often requires procedures spanning a variety of topics, including incident management, identity and access management, change and release management, patch management and project planning. These processes must be defined and deployed consistently across the enterprise. Without them, or if they are deployed haphazardly, the enterprise may be at risk of inconsistent management and results, lack of governance and reporting and failure to ensure compliance with regulations.

Ensuring consistent and sustainable operations is contingent on having repeatable and defined procedures, which ensure ease of overall continuity, reducing interruptions/disruptions from staff turnover. Well-defined procedures provide a foundation for the enterprise to achieve their goals and objectives.

CHAPTER 2— IT RISK ASSESSMENT

Industry Trends

Marketing staff is concerned with changes in trends, and a failure to see a changing trend can result in an enterprise losing a substantial percentage of its market share within a very short time period. This also applies to risk management functions. A failure by the IT department to adapt to or support a new business model may result in substantial losses to the enterprise. For example, when some telecommunications providers switched to per second instead of per minute billing, it forced all of the other telecommunications providers to change their systems quickly or lose a large percentage of their customers.

A risk practitioner should assess the maturity of the IT department and the enterprise as a whole toward monitoring and adapting to new market trends. A lack of flexibility or poor communication between the business units and IT could be a risk factor.

Forecasting Risk

Many times, a risk is realized as the result of preexisting or predisposing conditions that have been present for an extended period prior to their exploitation. Risk events may also cascade from one another, where one seemingly minor problem or vulnerability serves as a trigger that leads to a serious problem such as a prolonged outage. The risk practitioner should consider this when assessing risk and look for instances in which factors that appear innocuous individually, collectively may trigger an event.

Previous incidents, audit reports and failures affecting both internal and external parties provide the risk practitioner with valuable insight. Each event offers a wealth of information on what may have gone wrong and how to better protect systems, detect incidents and respond more effectively in the future. The greatest threat to the incident response process is a failure to learn from past events and repeat the same errors in the future. Investigations of past events may also validate or refine the calculation of risk impact, but the risk practitioner is cautioned to remember that no two incidents are identical and that the impact of a future event may differ substantially from a past event with similar characteristics.

2.2 Threat Modeling and Threat Landscape

Threats can be external or internal, intentional or unintentional. They may be caused by natural events or political, economic or competitive factors. Threats always exist and are typically beyond the direct control of the risk practitioner or asset owner. Not all conceivable threats need to be considered by every enterprise. For example, an enterprise that operates in a region with a seismic rating of zero does not have to document exposure to volcanoes or earthquakes. However, it is important to identify the various types of threats that do apply and may be used to compromise systems or otherwise affect the enterprise.

Threats may be divided into multiple categories, including:[1]

- Physical
- Natural events
- Loss of essential services
- Disturbance due to radiation
- Compromise or disclosures of information
- Technical failures
- Improperly-defined business logic
- Unauthorized actions
- Compromise of functions

CHAPTER 2— IT RISK ASSESSMENT

The risk practitioner should document all the threats that may apply to the systems and business processes under review. This includes using similar techniques to risk identification but should also include examining the cause of past failures, audit reports, media reports, information from national CERTs, data from security vendors and communication with internal groups.

Threats may be the result of accidental actions, intentional/deliberate actions or natural events. Threats may originate from either internal or external sources, and actual attacks may leverage a combination of both internal and external sources. Where threats may be directed by individual actors, the risk practitioner should know and remain aware that such threat actors tend to be imaginative, creative and determined and will explore new methods and avenues of attack. To counter threat actors, the risk practitioner must also be determined and creative, and seek to discover as many threats as possible. An unidentified threat is one for which the organization is more likely to be unprepared and vulnerable than a threat that is well documented.

Sources for information regarding threats are listed in **figure 2.6.**

Figure 2.6—Sources for Threat Information	
Service providers	Insurance companies
Threat monitoring agencies	Product vendors
Security companies	Government publications
Audits	Assessments
Management	Users
Business continuity	Human resources
Finance	Media

2.2.1 Internal Threats

Although many organizations label their employees and other staff to be their best assets, employees may be unhappy if they are inadequately trained, treated poorly or not given enough time to do their jobs properly. Disillusionment and resentment can lead to a higher risk of errors, negligence and more conscious actions such as theft. Key personnel also may be drawn to another company and leave serious gaps in the knowledge and skills needed to operate systems effectively.

Employees are the cause of a significant number of business impacts, which can be intentional and unintentional. A disgruntled employee may intentionally compromise systems or release data that expose the enterprise to legal or reputational risk. Employees may be coerced to disclose trade secrets for ideological or economic reasons or may be under duress. Many employees have access to systems and data that far exceeds their actual job requirements, which can be exploited in an attack. The solution to this problem is proper provisioning access to ensure that principles such as need-to-know and least privilege are introduced, but on its own, it is an incomplete solution. Any system has trusted insiders, and one of them choosing to violate trust is difficult to either predict or prevent.

The typical malicious insider is a current or former employee, contractor or other business partner who has or had authorized access to an enterprise's network, system or data and intentionally intercepted (exfiltrated), interrupted, modified or fabricated data on the enterprise's information systems. The first step in addressing personnel threats is to start with the hiring process and to review the qualifications and attitude of prospective employees. Employment candidates may have submitted incorrect information on job applications and claimed education, certification or experience that they did not actually possess. At the time of hiring, the employee should be required to sign a nondisclosure agreement (NDA) and be advised of the ethics and policies of the enterprise, and a review of references and performance of background checks may be worthwhile where permitted by law. See Section 1.12 Professional Ethics of Risk Management for more information.

Throughout employment, employees should be reminded of organizational policies and their responsibilities through awareness sessions and regular management reviews. One of the best employee-based controls is to interact with employees to understand any frustrations, complaints or issues that they may be facing and to seek to resolve those issues. During times of strike, layoffs, mergers, relocation and reorganization, an employee is more likely to be a threat. An employee who has been recently demoted or bypassed for a promotion is also a threat.

At the end of employment, an employee should return all organizational assets, including identification badges, equipment (e.g., laptops, mobile phones, access cards, etc.) and uniforms so that they cannot use those to gain unauthorized access in the future. In addition, systems, network and facility access should be revoked immediately prior to the employee's departure to minimize the potential for crimes of opportunity.

Internal threats can also originate from third parties that already operate within the internal environment of an enterprise. The risk practitioner should also consider vendor staff, business process outsourcing staff or any other external individuals that operate from within the enterprise. Although they can be considered as external threats, they usually have some internal access points, and higher knowledge and capabilities to potentially initiate a threat from within an enterprise.

2.2.2 External Threats

In a networked environment where data are stored or processed offsite or with third-party service providers, threats to information systems from outside the enterprise can originate from anywhere and may take a number of forms, including, but not limited to, the following:

- Espionage
- Theft
- Sabotage
- Terrorism
- Criminal acts
- Software errors
- Hardware flaws
- Mechanical failures
- Lost assets
- Data corruption
- Facility flaws (freezing pipe/pipe burst)
- Fire
- Supply chain interruption
- Industrial accidents
- Disease (epidemic)
- Seismic activity
- Flooding
- Power surge/utility failure
- Severe storms

Natural events such as a flood, storm, earthquake or tornado are unpredictable and may be extremely disruptive and damaging. The use of governmental data and weather monitoring services may identify the threats associated with natural events and allow the risk practitioner to take necessary steps to be prepared.

The range of external threat actors include criminals, hacktivists, corporate spies, thieves or advanced persistent threats (APT), each having different methods, motivations and capabilities. Where nation-state threat actors will be skilled and determined to break into systems for military or economic purposes, hacktivists will have a different set of skills and motivations to bare in an attempt to publicly humiliate or shame an organization. The term APT refers to advanced, highly-skilled attackers that are determined (persistent) in their attempts to exploit systems and networks. They pose increased skills and the effectiveness of the tools they possess makes the risk of compromise significant. APTs may be sponsored by governments, organized crime or competitors.

Most breaches are the result of targets of opportunity, not determined attacks. As seen in the annual reports from Verizon[2] and other organizations, many enterprises are breached because they were discovered to be easy targets and threat actors took advantage of their vulnerabilities.

2.2.3 Emerging Threats

Indications of emerging threats may include unusual activity on a system, repeated alarms, degraded system or network performance, or new or excessive activity in logs. In many cases, compromised organizations have evidence of emergent threats in their logs well in advance of the actual compromise, but the evidence is either not acted on or noticed. Lack of effective monitoring and response, when combined with a threat, is a combination that can lead to a breach.

Most technologies are built with an emphasis on function and purpose without due consideration for the security implications. As a result, new technology tends to be a source of new vulnerabilities and may even itself be a threat actor within an information system. The risk practitioner must be alert to the emergence of new technologies and prepare for their introduction into the enterprise, particularly if these technologies promise cost savings or competitive advantage. Internet of Things (IoT) is an example of a revolution in how enterprises view technology assets, one whose risk is self-evident but that has tempted a wide variety of enterprises by promising to greatly reduce the cost of both initial procurement of IT assets and the rate at which they need to be refreshed. A strategy that emphasizes rejection of new technology is unlikely to remain in place long beyond the point at which something gains executive sponsorship.

See section 3.5 Management of Emerging Risk for more information.

2.2.4 Additional Sources for Threat Information

Additional sources for threat information are discussed in the following sections.

Conducting Interviews

Gathering information from staff is a valuable method of gaining insight into the business from those closest to the processes and most likely to understand their fundamental workings. However, interviews pose certain challenges. One is that many people want to be seen as essential participants in the mission of the enterprise, which can lead to exaggeration of their own importance and that of their teams or departments. Another is that people may not fully understand the overall business process or the dependencies between their department and other departments. In certain cases, such as where there is demonstrable negligence or wrongdoing, someone may intentionally provide incorrect information.

The risk practitioner can improve the results of an interview producing useful information by adhering to the following good practices:
- Designate a specific time period and do not exceed that time without mutual agreement.

- When a manager is told that a staff member will be needed for 45 minutes, they should not discover that the interview lasted 90 minutes.
- Know as much as possible about the business process in advance of the interview, which reduces time spent on general explanations of core business functions.
  - Obtain and review documentation such as process maps, standard operating procedures, the results of impact assessments and network topologies before the interview.
- Prepare questions and provide them to the interviewee in advance so that they can bring any supporting documentation, reports or data that may be necessary.
- Conduct interviews with senior leaders in order to ensure a thorough understanding of the enterprise, including every aspect of each business operation.
  - Senior leaders may include board members, administrators, critical third-party service providers, customers, suppliers and managers.
- Encourage interviewees to be open about challenges they face and risk that concerns them along with any potential missed opportunities or problems associated with their current processes, systems and services/products.
- Avoid setting incorrect expectations regarding confidentiality of interview answers.
  - People may worry about the repercussions of discussing flaws or missed opportunities.
  - Promise confidentiality only if it will actually be maintained.

One important detail to obtain during an interview is the level of impact that previous incidents have had on the enterprise, including how the incident was handled, results of post-incident review and root cause analysis and current status of any noted remediation activities from prior incidents, including whether the recommendations from the previous incident were addressed and resolved. The risk practitioner should be aware that the people made available for interviews may not be the correct people to answer risk-related questions. When scheduling or other conflicts impede the risk-management process, the risk practitioner may need to turn to the executive sponsor of the program to facilitate access to the right people in a timely manner.

Media Reports

Media sources—such as newspapers, television, articles and social media—can be useful sources of information about threats or incidents but should be considered with a healthy degree of professional skepticism and critical thought. The risk practitioner should alert those with authority when a media communication could impact the enterprise or its employees, customers or business partners and should be prepared to advise on how to respond to a threat mentioned in the media that could affect the enterprise or a product or service used by the company's internal or external stakeholders. All stakeholders need to be educated on potential risk alluded to by the news feed by the appropriate department in the enterprise.

Observation

Watching a process as an independent observer may highlight issues that are unable to be seen as clearly by those performing the operations and can help to identify situations in which documented processes are not being followed or where processes are found to be lacking. Evaluating outcomes, start to finish, of exercises and real-world activities from the perspective of an observer is particularly valuable for the risk practitioner, who should first consider what is actually happening and ensure that it is reflected in workflows; later, these workflows can be compared to stated procedures and the expected outcomes from the process.

CHAPTER 2— IT RISK ASSESSMENT

Besides making observations, the risk practitioner should conduct or review the business process mapping to identify out-of-band services and new ways of working created during periods of business reorganization that are not captured in the previous operating models. Shadow IT should also be considered.

Logs

Logs are powerful sources of threats. These logs, when ingested in security information and event management tools, can be used to correlate events from different sources to identify a threat or an incident.

Self-assessments

Many enterprises encourage direct manager involvement in monitoring risk and control effectiveness within their areas of responsibility. Local managers typically have insight into the behavior of their staff and routine operations, so they are well-suited to evaluate compliance with procedures, recurring problems, risk trends and vulnerabilities. Properly conducted self-assessments can provide the risk team and senior management with regular reporting of risk and assurance of control effectiveness and may also reduce the need for more intense audits or system reauthorization.

Third-party Assurance

Third-party performance of reviews, audits, compliance verification and analysis of the enterprise, its processes, incidents, logs and threat environment can be a valuable source of information about risk, control effectiveness (design and operational efficacy), vulnerabilities and compliance. The enterprise may also benefit from the expertise, objectivity and credibility of a third-party organization, whose assurance that the organization is managing risk in accordance with senior management's direction and is compliant with standards and good practices may be seen as more credible than the results of a self-assessment conducted by staff within the organization who may be biased to ensure performance goals are being achieved or otherwise professionally or personally influenced.

User Feedback

System users are in an excellent position to know the system and its shortcomings, including potential vulnerabilities. Interacting with users and learning their issues and concerns regarding the system may indicate where security controls could be circumvented for convenience or improved efficiency. For instance, if users are regularly exporting data from a system in order to create or customize reports or perform analysis using basic office applications, the risk practitioner should verify whether such practices are formally recognized by procedure or have been adopted on an *ad hoc* basis, in which case they may present unaccepted risk.

Vendor Reports

There are many sources for information on current threats and vulnerabilities, new types of malware or emerging attack methods. Many government-sponsored CERTs and security vendors provide free or subscription-based reports and analysis, and these can be valuable sources of information for the risk practitioner.

A renewed focus on the global IT supply chain posed a challenging risk that cannot be ignored, both in delivery resiliency and security. Many governments are controlling the source of product imports and exports and associated applications. Therefore, enterprises need to ensure they are assessing risk associated with this area, and vendor reports can provide necessary information.

2.2.5 Threat, Misuse and Abuse-Case Modeling

The mindset and approach of the adversary and the extent to which motivation and skill level of the attacker are important to understand and useful in determining the real existence of a threat. Threat modeling examines the nature of the threat and potential threat scenarios, which are valuable tools that can greatly assist in risk assessment. **Figure 2.7** shows the components of a threat scenario.

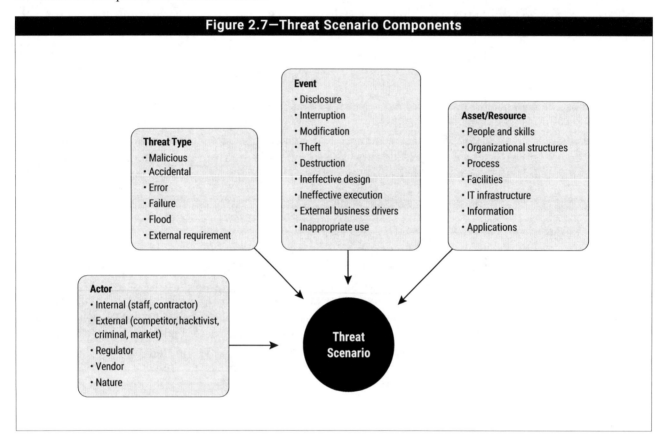

At a high-level, threat modeling is done by mapping the potential methods, approaches, steps and techniques used by an adversary to perpetrate an attack. The threat agent will often try different tools, probe for vulnerabilities and try both technical and nontechnical approaches while seeking to compromise a system. The risk practitioner must think of as many of these methods and approaches as possible so that adequate controls can be designed to meet the possible threats. Threat modeling helps the risk practitioner, systems designers, developers and operators to build systems with attention to defensive controls, built-in security features and proper placement within a strategy of overlapping defenses (defense in depth) that can deter or prevent system failures.

Threat modeling, misuse and abuse-case modeling differ from standard use-case modeling. Where use-case modeling examines how a system will function to deliver value to its users, misuse-case modeling looks at a variety of possible errors, mistakes, unintentional deviations from expected user behavior that a system may experience. Abuse-case modeling examines security and privacy interactions of a system, user, data and stakeholders. Abuse cases consider the complete interaction between a system and one or more actors, where the results of the interaction are harmful to the system, users or one of the stakeholders in the system.

Understanding misuse and abuse-cases helps to ensure that a system is designed and built with resiliency and has the ability to handle errors, misuse and abuse.

Threat attribution is the process used in an attempt to identify the threat actors behind an attack and their motivations. This typically relies heavily on the use of forensic analysis to find evidence, cataloging known indicators of compromise (IOCs) and indicators of attack (IOAs), in order to associate an attack with a specific threat actor. While not impossible, threat attribution can be difficult, and must be balanced with the strategy, goals and objectives of the organization. The purpose of establishing capabilities such as "cyber attribution" have to be fully considered as to the value and benefit provided to the organization, as the accuracy and confidence associated with attribution will vary depending on quality of the data available and can require a significant investment.

Threat Modeling

Threat modeling is an important part of the risk analysis process as it is important to identify the variety of actors associated with a given scenario (not necessarily with the same level of rigor as attribution—though positive attribution can accelerate the process).

It is important to understand which controls are successful at defending against the attacks or errors associated with various threat communities. In some cases, controls may address threats from more than one source. For example, the same control that addresses nation-state-sponsored attacks may also defend against cybercriminals. Overlapping controls in this sense should be noted in the threat model.

The answers to these questions can be collected in a threat profile and maintained for each threat community identified by the enterprise. The profile can then be correlated to the appropriate threat identified and subsequent risk scenarios to gain a clearer picture of end-to-end enterprise risk.

Threats can come from both outside or within enterprises, and they can have devastating consequences if realized. Attacks can disable systems entirely or lead to the leaking of sensitive information, which would diminish consumer trust in the system provider. To prevent threats from taking advantage of system flaws, various threat-modeling methods can steer appropriate response and help make informed decisions around appropriate safeguards and countermeasures needed to protect an asset.

In general, threat modeling will:
- Create an abstraction of the system
- Profile attackers, including goals, methods and attacker skills
- Create a catalog of potential threats

Additionally, it will provide high-level answers to basic questions, such as:
- How often will a threat actor encounter assets at the vendor location?
- What is the danger of discovery posed to that threat actor while it is attempting to compromise those assets?
- What perceived value may enterprise data at the third party hold for an adversary?
- What skills does an adversary need to succeed in compromising vendor systems and accessing enterprise data?
- How much time does an adversary need to compromise systems and access data?
- What resources and materials does the adversary need?
- What level of effort is required overall from the threat actor to compromise the vendor?

The answers to these questions can be collected in a threat profile and maintained for each threat community identified by the enterprise. The profile can be correlated to risk statements, third parties and business processes to gain a clearer picture of end-to-end enterprise risk.

Several threat-modeling methods can be combined to create a more robust and well-rounded view of potential threats. Not all methods are comprehensive; some are abstract, and others are people-centric. Some methods focus specifically on risk or privacy concerns. **Figure 2.8** provides a list of threat model methods.

Figure 2.8—Threat Model Methods		
Model	Benefits	Background
STRIDE (Spoofing identity, Tampering with data, Repudiation, Information disclosure, Denial of service, Elevation of privilege)	<ul><li>Helps identify relevant mitigating techniques</li><li>Most mature</li><li>Easy to use but time consuming</li></ul>	Invented in 1999 and adopted by Microsoft in 2002, STRIDE is currently the most mature threat-modeling method. STRIDE evaluates the system detail design. It models the in-place system. By building data-flow diagrams (DFDs), STRIDE is used to identify system entities, events and the boundaries of the system.
PASTA (Process for Attack Simulation and Threat Analysis)	<ul><li>Helps identify relevant mitigating techniques</li><li>Directly contributes to risk management</li><li>Encourages collaboration among stakeholders</li><li>Contains built-in prioritization of threat mitigation</li><li>Laborious, but has rich documentation</li></ul>	PASTA is a risk-centric threat-modeling framework developed in 2012. It contains seven stages, each with multiple activities. PASTA aims to bring business objectives and technical requirements together.
LINDDUN (Linkability, Identifiability, Nonrepudiation, Detectability, Disclosure of information, Unawareness, Noncompliance)	<ul><li>Helps identify relevant mitigating techniques</li><li>Contains built-in prioritization of threat mitigation</li><li>Can be labor intensive and time consuming</li></ul>	LINDDUN focuses on privacy concerns and can be used for data security. This method starts with a DFD of the system that defines the system's data flows, data stores, processes and external entities. LINDDUN can be used to help identify a threat's applicability to the system and build threat trees through the analysis of model elements from the point of view of threat categories.
Attack trees	<ul><li>Helps identify relevant mitigating techniques</li><li>Has consistent results when repeated</li><li>Easy to use if a thorough understanding of the system is already in place</li></ul>	Attack trees are diagrams that depict attacks on a system in tree form. The tree root is the goal for the attack, and the leaves are ways to achieve that goal. Each goal is represented as a separate tree. Thus, the system threat analysis produces a set of attack trees.
Persona non Grata (PnG)	<ul><li>Helps identify relevant mitigating techniques</li><li>Directly contributes to risk management</li><li>Has consistent results when repeated</li><li>Tends to detect only some subsets of threats</li></ul>	PnG focuses on the motivations and skills of human attackers. It characterizes users as archetypes that can misuse the system and forces analysts to view the system from an unintended-use point of view. The idea is to introduce a technical expert to a potential attacker of the system and examine the attacker's skills, motivations, and goals. This analysis helps the expert understand the system's vulnerabilities from the point of view of an attacker. PnG tends to work very well with the Agile approach, which uses personas.

Figure 2.8—Threat Model Methods (cont.)

Model	Benefits	Background
Trike	• Helps identify relevant mitigating techniques • Directly contributes to risk management • Contains built-in prioritization of threat mitigation • Encourages collaboration amount stakeholders • Has automated components • Has vague, insufficient documentation	Trike was created as a security audit framework that uses threat modeling as a technique. It looks at threat modeling from a risk-management and defensive perspective. As with many other methods, Trike starts with defining a system. The analyst builds a requirement model by enumerating and understanding the system's actors, assets, intended actions and rules. This step creates an actor-asset-action matrix in which the columns represent assets, and the rows represent actors.
VAST (Visual, Agile and Simple Threat)	• Helps identify relevant mitigating techniques • Directly contributes to risk management • Contains built-in prioritization of threat mitigation • Encourages collaboration among stakeholders • Has consistent results when repeated • Has automated components • Is explicitly designed to be scalable • Has little publicly available documentation	VAST modeling method is based on ThreatModeler, an automated threat-modeling platform. Its scalability and usability allow it to be adopted in large enterprises throughout the entire infrastructure to produce actionable and reliable results for different stakeholders. Recognizing differences in operations and concerns among development and infrastructure teams, VAST requires creating two types of models: • Application threat models, which use process-flow diagrams, representing the architectural point of view. • Operational threat models, which are created from an attacker point of view based on DFDs. This approach allows for the integration of VAST into the organization's development and DevOps life cycles.

Threat modeling is similar to misuse and abuse modeling, in that it examines the different threats which may result in a system being attacked and used for a purpose for which the system was never intended. An example of this is the "ping of death" attack. The Internet Control Message Protocol (ICMP) was designed as a tool for system and network administrators to test network connectivity, but by altering the size of an ICMP packet, an attacker is able to build an attack tool that could disable target system. Similar threats arising from intentional misuse have used the Network Time Protocol (NTP) and domain name system (DNS) services as attack platforms.

2.3 Vulnerability and Control Deficiency Analysis

Vulnerabilities are weaknesses, gaps or holes in an enterprise's people, processes or technologies that provide an opportunity for a threat actor to exploit, which create consequences that may impact the enterprise.

The *National Institute of Standards and Technology (NIST) Special Publication 800-30 Revision 1: Guide to Conducting Risk Assessments* provides a list of vulnerabilities to consider along with "predisposing conditions" that may lead to the rapid or unpredictable emergence of new vulnerabilities. Many vulnerabilities are conditions that exist in systems and must be identified so that they can be addressed. The purpose of vulnerability identification is to find the problems before they are found by an adversary and exploited. This is why an enterprise should conduct regular vulnerability assessments and penetration tests to identify, validate and classify its vulnerabilities. Where vulnerabilities exist, there is a potential for risk.

CHAPTER 2— IT RISK ASSESSMENT

> Note: Other sources of vulnerabilities include:
> - National Vulnerability Database at nvd.nist.gov
> - Common Weakness Enumeration at nvd.nist.gov/vuln/categories
> - Common Vulnerabilities and Exposures at cve.mitre.org
> - Exploit Database at exploit-db.com
> - The Community Driven Vulnerability Database at vulndb.org
> - Vulnerability Database Catalog at www.first.org/global/sigs/vrdx/vdb-catalog
> - Packet Storm Security at packetstormsecurity.com
> - Rapid7's Vulnerability and Exploit Database at rapid7.com/db
> - CXSecurity Free Vulnerability Database at cxsecurity.com/exploit/
> - Vulnerability Lab at vulnerability-lab.com
> - 0day.today at www-.0day.today
> - Harmonized Threat and Risk Assessment at www.cse-cst.gc.ca
> - Open Web Application Security Project (OWASP) at www.owasp.org

2.3.1 Sources of Vulnerabilities

Vulnerabilities can be found in most places within and external to the enterprise. Sources of vulnerabilities are discussed in the following sections.

Network Vulnerabilities

Network vulnerabilities are often related to misconfiguration of equipment, poor architecture or traffic interception. Misconfiguration is a common problem with network equipment that is not properly installed, operated or maintained. Any open services are a potential attack vector that can be exploited by an attacker, so network equipment should be hardened by disabling any unneeded services, ports or protocols.

Traditionally, networked systems have been limited to devices whose primary purpose was to provide computing power, either for direct processing of data or to regulate the flow of data within the network. However, as the cost of networking technology drops in terms of price, power and physical size, the ability to transmit and receive data is being incorporated into a growing number of devices whose primary purposes are not of a data processing or regulatory nature, such as home appliances and automobiles. These devices create new vulnerabilities for the network. IoT is driving both a tremendous opportunity for increasing convenience and a target for threat agents. As concepts such as edge computing are introduced to alleviate bandwidth constraints, they are also carry with them a move away from traditional security models normally associated with data centers and push capabilities to smart technologies and mobile phones ("the edge") to perform tasks once only performed within the confines of the datacenter. Therefore, the risk practitioner should ensure they are knowledgeable on emerging technologies and their implications for future use.

Physical Access

Physical security goes far beyond theft. Threat agents that are able to gain physical access to systems have the potential to bypass nearly every other type of control. With access to server rooms, network cabling, information systems equipment and buildings, an attacker can circumvent passwords, install "skimmers" to intercept data communications and take logical ownership of systems or devices. With more than 20 percent of incidents related to misuse of systems by insiders who may already have an advantage in terms of gaining physical access, the need for strong physical security is clear and growing. While there has been a decline over the last few years, physical security concerns are still very valid and are a reliable method for gaining initial access to an environment (**figure 2.9**).

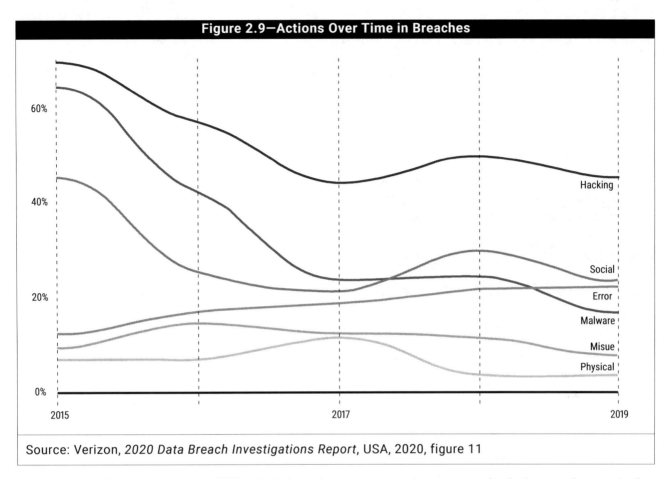

Figure 2.9—Actions Over Time in Breaches

Source: Verizon, *2020 Data Breach Investigations Report*, USA, 2020, figure 11

Testing for physical security vulnerabilities include testing access control systems, testing locks, security guards, fire suppression systems, heating ventilation and air conditioning controls, lighting, cameras, motion sensors and physical entrance.

Applications and Web-facing Services

Applications in general, web applications in particular, are among the most common entry points currently used by attackers. Many applications are written to support business function without due regard to the security requirements of the applications and may be vulnerable to buffer overflows, logic flaws, injection attacks, bugs, incorrect control over user access and many other common vulnerabilities. The risk practitioner can use tools from the Open Web Application Security Project (OWASP)[3] to test web-facing applications for well-known and documented vulnerabilities whose presence can lead to data compromise or system failure.

Applications may also be vulnerable on account of poor architecture. Applications that process payment card data or other sensitive information or store those data in an insecure location, such as in a demilitarized zone (DMZ), may come under repeat or even continuous attack, increasing the likelihood that an attacker will find a vulnerability to exploit. The risk practitioner should understand the network architecture and question decisions that appear to create unnecessary exposure.

Utilities

Information systems rely on controlled environmental conditions, including clean and steady power and controls over humidity and temperature. The risk practitioner should ensure that the enterprise is prepared for power failure or

CHAPTER 2— IT RISK ASSESSMENT

other environmental conditions in order to avert conditions that may lead to system failure. Having an uninterruptible power supply (UPS), backup generators and surge protectors can protect equipment from damage or failure, but these systems require more than an initial investment. For instance, a UPS must be validated to ensure it has adequate power to run critical systems until a backup generator can come online and then tested periodically, while generators must be checked for fuel and mechanical function.

Supply Chain

Many enterprises rely on products, raw materials and supplies that originate from various locations around the world, and any interruption in the supply chain may affect their ability to function. For example, a shortage in supply of fuel to an airport would impact all flight operations. Reliance upon a supply chain should be documented even if it is understood that the risk is one that is accepted by senior management.

Equipment

As equipment ages, it becomes less efficient, effective and potentially unable to support business functions. Equipment is often provided at the time of production with a mean time between failure (MTBF) rating that indicates its anticipated life span and when it should be scheduled for removal or replacement.

There are multiple threats that older equipment can pose to enterprises, and each needs to be investigated and considered in the full context of the business value provided and benefits yielded.

Careful consideration must be made to ensure that the risk practitioner does not assess solely in context of a technology-centric view. The risk practitioner needs to take steps to fully understand the business context, value generated by and benefit such a system brings to the enterprise in order to properly assess associated risk. While recommending the replacement of end-of-life equipment may address risk with technology debt, it also introduces additional risk associated with business operations, existing processes and disruptions. These factors need to be considered when assessing equipment.

There may be instances where core business processes are built around legacy systems. In these cases, the cost and disruption to the business may exceed the benefit of bringing in a new system.

Cloud Computing

Outsourcing of application hosting and data processing has been growing in popularity for decades. However, the growing availability and declining costs in recent years of both high-speed connectivity and systems capable of hosting virtual machines have transformed the practice. Services hosted in third-party data centers now have astonishing redundancy, automatic failover and SLAs that guarantee uptime in excess of 99.9 percent—all through an on-demand pricing structure that can undercut in-house operations by an order of magnitude. These characteristics make the use of cloud-based services very attractive to many enterprises. There are several cloud-based service models, as seen in **figure 2.10**.

Figure 2.10—Cloud Deployment Models	
Private cloud	Operated solely for an enterprise
	May be managed by the enterprise or a third party
	May exist on- or off-premises
Public cloud	Made available to the general public or a large industry group
	Owned by an organization selling cloud services

CHAPTER 2— IT RISK ASSESSMENT

Figure 2.10—Cloud Deployment Models *(cont.)*	
Community cloud	Shared by several enterprises
	Supports a specific community that has a shared mission or interest
	May be managed by the enterprises or a third party
	May reside on- or off-premises
Hybrid cloud	A composition of two or more clouds (private, community or public) that remain unique entities, but are bound together by standardized or proprietary technology that enables data and application portability (e.g., cloud bursting for load balancing between clouds)
Source: ISACA, *IT Control Objectives for Cloud Computing: Controls and Assurance in the Cloud*, USA, 2011, figure 1.3	

Cloud computing offers many business advantages, both the organization and the risk practitioner should be conscious that the outsourcing of data processing does not remove the liability of the outsourcing organization to ensure proper data protection. Also, while larger data centers that are established to serve hundreds or thousands of customers are typically better protected than stand-alone operations, the ability of the outsourcing organization to see what goes on inside the company that hosts its data center may be very limited. When one considers that the cloud data centers are large, visible and tempting targets for well-funded or sufficiently motivated threat actors, it is clear that cloud computing represents risk and opportunity. The decisions of whether to leverage cloud services, which systems should be transferred and which might be better retained within the control of the organization should be made by senior management. In order to assist them in making this decision, the risk practitioner should provide senior management with the necessary information so they can have a thorough understanding of the risk involved.

Cloud computing service models, deployment models and characteristics are shown in **figures 2.11 and 2.12**. The risk of choosing the right services (SaaS, PaaS, IaaS) and deployment models as shown in **figure 2.10** must be carefully deliberated with the business and legal requirements in mind. A clear understanding of the shared responsibility between the customer and the Cloud Service Provider (CSP) is needed as this will help to alleviate many impacts to the business.

Figure 2.11— Cloud Computing Service Models		
Service Model	Definition	To Be Considered
Infrastructure as a Service (IaaS)	IaaS is the capability to provision processing, storage, networks and other fundamental computing resources, offering the customer the ability to deploy and run arbitrary software that may include operating systems and applications. IaaS puts these IT operations into the hands of a third party.	Options to minimize the impact if the cloud provider has a service interruption
Platform as a Service (PaaS)	PaaS is the capability to deploy onto the cloud infrastructure customer-created or acquired applications using programming languages and tools supported by the provider.	• Availability • Confidentiality • Privacy and legal liability in the event of a security breach, as databases housing sensitive information will be hosted offsite • Data ownership • Concerns around e-discovery

CHAPTER 2— IT RISK ASSESSMENT

Figure 2.11– Cloud Computing Service Models *(cont.)*		
Service Model	Definition	To Be Considered
Software as a Service (SaaS)	Capability to use the provider's applications running on cloud infrastructure. The applications are accessible from various client devices through a thin client interface such as a web browser (e.g., web-based email).	• Who owns the applications? • Where do the applications reside?

Source: ISACA, *Cloud Computing: Business Benefits With Security, Governance and Assurance Perspectives*, USA, 2009, www.isaca.org/Knowledge-Center/Research/ResearchDeliverables/Pages/Cloud-Computing-Business-Benefits-With-Security-Governance-and-Assurance-Perspective.aspx

Figure 2.12–Cloud Computing Essential Characteristics	
Characteristic	Definition
On-demand self-service	The cloud provider should have the ability to automatically provision computing capabilities, such as server and network storage, as needed without requiring human interaction with each service's provider.
Broad network access	According to NIST, the cloud network should be accessible anywhere, by almost any device (e.g., smart phone, laptop, mobile devices).
Resource pooling	The provider's computing resources are pooled to serve multiple customers using a multitenant model, with different physical and virtual resources dynamically assigned and reassigned according to demand. There is a sense of location independence. The customer generally has no control or knowledge over the exact location of the provided resources. However, he/she may be able to specify location at a higher level of abstraction (e.g., country, region or data center). Examples of resources include storage, processing, memory, network bandwidth and virtual machines.
Rapid elasticity	Capabilities can be rapidly and elastically provisioned, in many cases automatically, to scale out quickly and rapidly released to scale in quickly. To the customer, the capabilities available for provisioning often appear to be unlimited and can be purchased in any quantity at any time.
Measured service	Cloud systems automatically control and optimize resource use by leveraging a metering capability (e.g., storage, processing, bandwidth and active user accounts). Resource usage can be monitored, controlled and reported, providing transparency for both the provider and customer of the utilized service.

Source: ISACA, *Cloud Computing: Business Benefits With Security, Governance and Assurance Perspectives*, USA, 2009, www.isaca.org/Knowledge-Center/Research/ResearchDeliverables/Pages/Cloud-Computing-Business-Benefits-With-Security-Governance-and-Assurance-Perspective.aspx

Migration to a cloud environment needs to be carefully weighed and considered in the context of the enterprise's strategic goals and objectives. Cloud service providers can appear to be an attractive option and they tend to make it both enticing and easy to engage with them. However, just as with any other business decision, there are some common considerations that require careful analysis and proper attention prior to migrating to a cloud-based environment.

- **Reduced visibility and control.** When moving to the cloud, enterprises will lose some visibility and control over operations and assets. While the cloud may sound like a cost-effective solution, enterprises need to carefully compare the benefits and value to be gained with costs. For example, network visibility is often a common regulatory, statutory and contractual requirement that the enterprise needs to demonstrate compliance with and there may be additional charges to gain the visibility required to address compliance requirements.
- **Self-service can lead to unauthorized use and abuse.** The ability to easily provision new services and spin up systems can open the enterprise up to new threats and liabilities. Think of how easy it could be for a department

to spin up a new system for a marketing campaign and client lists are exposed due to lack of proper configuration or having to absorb the costs for an administrator who stands up a cryptocurrency mining system and lets it run for a month.

- **Provider's privileged insiders**. The cloud provider's staff and administrators can abuse their authorized access to the enterprise's networks, systems and data and are uniquely positioned to cause damage or exfiltrate information.
- **Application programming interfaces (APIs) can be compromised.** An exposed service, such as a set of APIs with cloud management, can be abused and exploited like everything else that is connected to the internet.
- **Separation between multiple tenants fails.** The exploitation of a system, applications or other software that supports multitenant environments can lead to the failure to maintain proper separation among those tenants.
- **Data deletion is incomplete.** Reduced visibility into where their data are stored within the cloud and the impaired ability to ground truth verify the secure deletion of their data is not available.
- **Credentials can (and will) be stolen/compromised.** Just like all other credentials, if an attacker gains access to a cloud user's credentials, the attacker can have access to the services where they can provision additional resources, abscond with data, in addition to being able to attach the enterprise's assets.
- **Vendor lock-in.** Providers make the cloud very easy to use but getting the enterprise's data out of the environment is never a stress-free undertaking.
- **Increased complexity strains IT staff.** Moving to the cloud can introduce both simplification and complexities. Managing, integrating and operating in a cloud environment will most likely require the enterprise's existing IT staff to be trained in how to properly use the environment to realize the value and derive benefit from a cloud model. The enterprise's IT staff must have the ability to both manage, integrate, and maintain the migration of assets and data in the cloud as well as the current responsibilities for on-prem IT resources.
- **Insufficient due diligence.** Enterprises that move to the cloud may skip performing due diligence altogether due to the misunderstanding of how various external business drivers apply to the two separate entities. A common misconception that organizations have is the belief that if a cloud provider has a certification (e.g., SOC II report or PCI DSS report on compliance) that certification will then transfer to the customer's environment, but this is not the case.

Big Data

Advances in the capability to perform analysis of data from various sources of structured and unstructured data allow enterprises to make better business decisions and increase competitive advantage. This change in analytics capabilities dealing with big data can introduce technical and operational risk, and organizations should understand that risk can be incurred either through adoption or non-adoption of these capabilities.

The risk of adopting or not adopting big data analytics is described by ISACA the following way:[4]

Technical and operational risk should consider that certain data elements may be governed by regulatory or contractual requirements and that data elements may need to be centralized in one place (or at least be accessible centrally) so that the data can be analyzed. In some cases, this centralization can compound technical risk.

For example:

- **Amplified technical impact**—*If an unauthorized user were to gain access to centralized repositories, it puts the entirety of those data in jeopardy rather than a subset of the data.*
- **Privacy (data collection)**—*Analytics techniques can impact privacy; for example, individuals whose data are being analyzed may feel that revealed information about them is overly intrusive.*
- **Privacy (re-identification)**—*likewise, when data are aggregated, semianonymous information or information that is not individually identifiable information might become non-anonymous or identifiable in the process.*

CHAPTER 2— IT RISK ASSESSMENT

Additionally, the benefits of big data need to be weighed in consideration to the enterprise's maturity, governance and management of leveraging big data to realize value and gain benefit, instead of big data becoming a big risk.

Some of the challenges[5] enterprises commonly face associated with big data include the following:

- Data ownership is often not clearly defined due to the misconception that data management is technical work and, therefore, the responsibility of the IT department.
- Marketing or other departments can obtain the data needed by their business units without considering the costs or risk associated with using the data.
- Evolving regulatory requirements impact data governance. The number of data-related regulations is trending upward (e.g., California Consumer Privacy Act [CCPA], General Data Protection Regulation [GDPR] on privacy protection and China Cybersecurity Law on data localization). The lack of a clear data governance structure makes it harder to respond to emerging or changing international data laws.
- Enterprises often identify a need for workflow changes, which has an impact on enterprise architecture. Opportunities to standardize, interoperate and re-architect current processes and systems must be taken. Proper change management, data protection and executive sponsorship are necessary to make data governance a reality, particularly in heavily regulated environments.
- Siloed departmental and organizational structures result in disaggregated datasets and data analytics challenges.
- Data deluge can result in quality problems. Big data is of no value if it cannot be harnessed for better business decision making. Effective decisions cannot be made based on poor quality data. In addition, more complex and unstructured data are being introduced, which can strain data governance efforts.
- Lack of skilled staff to conduct data analysis creates a barrier to effective data governance. A good working knowledge of data analytics becomes critical to the success of an enterprise. The best resources should be able to identify hidden patterns and unknown correlations, draw conclusions, make inferences and predict future trends.

2.3.2 Gap Analysis

By documenting the desired state or condition of risk that management wants to reach and then carefully analyzing and evaluating the current condition of the organization, the risk practitioner can identify the existence of a risk gap and the scope of actions that may be needed to close the gap. Gap analysis can be used in iteration to devise deliverables, projects and milestones. This process is illustrated in **figure 2.13**.

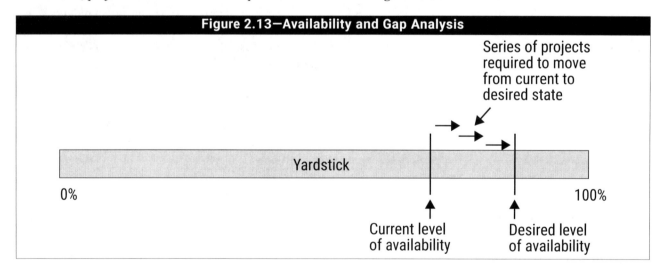

Figure 2.13—Availability and Gap Analysis

Using the gap analysis method to plan projects, along with milestones, can help the organization initiate and execute projects in an orderly and logical manner that considers their dependencies among each other.

To measure the current state, the risk practitioner may find it beneficial to identify one or more key indicators that relate to goals or performance (i.e., key goal indicators [KGIs] or key performance indicators [KPIs]), which may be internal to the organization to relate to identified standards, frameworks or good practices. Desired states are typically linked to risk (i.e., key risk indicators [KRIs]). KPIs and KRIs are discussed in chapter 3 Risk Response and Reporting.

2.3.3 Vulnerability Assessment and Penetration Testing

A vulnerability assessment is a careful examination of a target environment to discover any potential points of compromise or weakness. Vulnerabilities that may be identified by an assessment include:

- Network vulnerabilities
- Poor physical access controls (e.g., buildings, offices)
- Insecure applications
- Poorly designed or implemented web-facing services
- Disruption to utilities (e.g., power, telecommunications)
- Unreliable supply chain
- Untrained personnel
- Inefficient or ineffective processes (e.g., change control, incident handling)
- Poorly maintained or old equipment

Vulnerability assessment may be manual or automated. Automated tools have the ability to filter large amounts of data and can be used to examine logs; compile and analyze data from multiple sources (e.g., a security incident and event management [SIEM] tool); examine the functions of a program; and run test files or data against a tool such as a firewall or application. Where content is not easily quantified and requires judgment—including reviews of business processes, physical security and source code—a manual test is better (although automated tools for reviewing code continue to improve).

Vulnerability assessments may contain data that are not accurate, such as false positives that indicate a vulnerability where none exists. It is important to ensure qualified staff are executing and interpreting results from automated tools. Vulnerability assessments may also miss the identification of vulnerabilities that require a sequence of chained techniques to exploit. To validate the results of a vulnerability assessment, the organization may conduct a penetration test against a potential vulnerability or attack vector.

Penetration testing can use the same sort of tools as those used by a real adversary, which can help to establish the extent to which an identified vulnerability is a true weakness. There are different types of penetration tests that can be performed that are based upon the enterprise's appetite and end goals, which should dictate the approach that is employed in carrying out a penetration test.

Penetration tests may be conducted by either internal or external teams or hybrid, and rules vary from full knowledge of the environment to zero-knowledge tests in which the testing team has no knowledge of the environment being attacked. An expert penetration testing team will be creative and attempt several types of tests to achieve their goals. They may exploit multiple vulnerable attack vectors or only one that gains them access. The risk practitioner can use the report generated in the process of risk identification.

Careful consideration needs to be given to and planning is required to clearly define a penetration test engagement. Each penetration test is unique in nature and can include any number of activities, based on the adversary being simulated to the scope. The scope, time constraints, targets, exclusions and implicit authority to carry out these activities need to be clearly defined prior to having a penetration test.

CHAPTER 2— IT RISK ASSESSMENT

It needs to be clearly understood that the goals of vulnerability assessments and penetration tests differ in scope, nature and outcome. When performing a vulnerability assessment, the goal is the identification of any misconfigurations and/or vulnerabilities that may exist within a target system or application. Vulnerability assessments tend to be non-intrusive in nature in that they will profile the system and attempt to elicit information to determine if the system is vulnerable to an attack. There are typically two types of technical vulnerability assessments: credentialed and non-credentialed. Credentialed assessments will significantly reduce the number of potential false positives but require elevated permissions in order to run in order to understand the specific system configuration, version and patch levels applied to the system. Non-credentialed assessments will contain a much higher number of potential false positives as they are unable to confirm the specific configuration and patch levels applied to the system. Very little value may be gained, or benefit received in conducting a non-credentialed assessment in the grander scheme of security operations and risk management.

The purpose of a penetration test is to simulate an adversary, using intrusive methods in an attempt to circumvent controls and exploit potential vulnerabilities within an organization in order to action on objectives; the objectives are normally defined in advance of the testing and clearly stated (e.g., get to this system, exfiltrate this information, compromise this individual). The goal of a penetration test is not to identify every potential exploitable vulnerability, but rather to discover and document how an adversary might leverage the current state environment in order to be able to accomplish their mission.

False Positives and Zero-Day Exploits

The risk faced by enterprises consists of a combination of known and unknown threats, directed against the enterprise that have some combination of known and unknown vulnerabilities. A vulnerability assessment that identifies no vulnerabilities is not equal to the system being invulnerable in the absolute sense. It means only that the types of vulnerabilities that the assessment was intended to detect were not detected. Failure to detect a vulnerability may be the result of its absence, or it may be a false negative arising from misconfiguration of a tool or improper performance of a manual review. In the case of a zero-knowledge penetration test that fails to identify opportunities to exploit a system, it may be that the team was unlucky or lacked imagination, which may not be true of an outside attacker.

Even if no known vulnerabilities exist within the system, the system may still remain vulnerable to unknown vulnerabilities, more of which are discovered every day (and some of these are stored or sold for future use as zero-day exploits). The likelihood of an attack is often a component of external factors, such as the motivation of the attacker, as shown in **figure 2.14**.

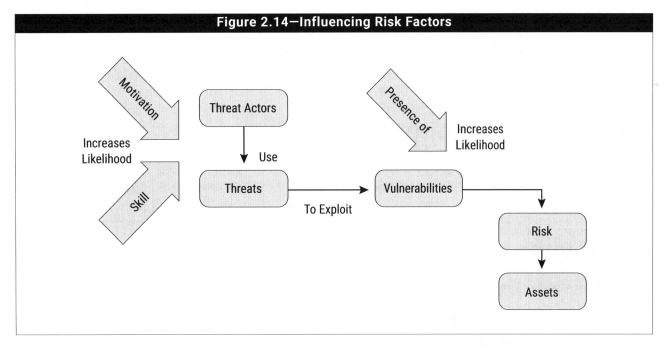

Figure 2.14—Influencing Risk Factors

When attempting to account for the variety of threat actors which exist, it is often beneficial to develop a threat actor profile[6] that can be used to help categorize the threat that a group of threat actors may pose to an enterprise's environment (**figure 2.15**). By developing a threat actor profile, the risk practitioner can reasonably account for certain elements associated with motivation and capabilities that could be brought to bear against the enterprise.

Figure 2.15—Threat Matrix

Threat Level	Threat Profile						
	Commitment			Resources			
					Knowledge		
	Intensity	Stealth	Time	Technical Personnel	Cyber	Kinetic	Access
1	H	H	Years to Decades	Hundreds	H	H	H
2	H	H	Years to Decades	Tens of Tens	M	H	M
3	H	H	Months to Years	Tens of Tens	H	M	M
4	M	H	Weeks to Months	Tens	H	M	M
5	H	M	Weeks to Months	Tens	M	M	M
6	M	M	Weeks to Months	Ones	M	M	L
7	M	M	Months to Years	Tens	L	L	L
8	L	L	Days to Weeks	Ones	L	L	L

Source: Duggan, David P.; Sherry R. Thomas; Cynthia K. K. Veitch; Laura Woodard; Categorizing Threat: Building and Using a Generic Threat Matrix, Sandia National Laboratories, USA, 2007, https://idart.sandia.gov/_assets/documents/SAND2007-5791_Categorizing-Threat_Generic-Threat-Matrix.pdf.

Given the combination of unknown threats and unknown vulnerabilities, it is difficult for the risk practitioner to provide a comprehensive estimate of the likelihood of a successful attack. Vulnerability assessments and penetration tests provide the risk practitioner with valuable information on which to partially estimate this likelihood, because:

- Although the presence of a vulnerability does not guarantee a corresponding threat, the all-hours nature of information systems and the rapid speed of processing makes it much more likely that an information system will come under an assortment of attacks in a shorter time than would be true of a physical system.
- A vulnerability known to an assessment tool is also knowable to threat agents, all but the most elite of whom tend to build their attacks to target common vulnerabilities.
- The presence of one or more known vulnerabilities—unless these have been previously identified and the risk is accepted by the enterprise for good reason—suggests a weakness in the overall security program.

The risk practitioner should ensure that senior management does not develop a false sense of security as a result of vulnerability assessments or penetration tests that fail to find vulnerabilities, but both forms of testing do provide valuable insight into the organization, processes, and its security posture.

2.3.4 Root Cause Analysis

Root cause analysis is a process of diagnosis to establish the origins of events so that lessons can be learned from consequences (typically errors and problems). The actions that an enterprise takes in response to risk are often based on the lessons learned from previous events. A prudent risk practitioner examines the root cause of an incident to discover the conditions and factors that led to the event, rather than reacting to the symptoms of the problem. These underlying conditions, which comprise the root (core) of a problem, are the conditions against which action must be taken in order to prevent the problem from recurring. For example, if a review of a business process finds that users are not compliant with the procedures and policies in place, the risk practitioner should examine why the users are not compliant before recommending enforcement of the procedure. It may be that the procedure is outdated, flawed or unworkable when aligned with the objectives of the enterprise, in which case it is the procedure and not the behavior that should be corrected. Root cause analysis examines the reasons why a problem exists or a breach has occurred and seeks to identify and resolve these underlying issues.

One implementation of root cause analysis is a pre-mortem—a facilitated workshop where the group is told to pretend that the project has failed and discuss why it has failed. When correctly facilitated, the simple but crucial cognitive impact of answering the question in this fashion (instead of why it *might* fail) can produce insightful, collaborative and valuable perspectives on risk.

In many cases, a risk event may be the result of coinciding events—several issues that act in combination to create what appears to be a single result. The risk practitioner can use root cause analysis as a means of identifying coinciding events, which cannot be traced to a single common cause. In this case, the solution is more than just a single response activity or implementation of a single control.

2.4 Risk Scenario Development

A risk scenario is a description of a possible threat event whose occurrence will have an uncertain impact on the achievement of the enterprise's objectives, which may be positive or negative. The development of risk scenarios provides a way of conceptualizing risk that can aid in the process of risk identification. Scenarios are also used to document risk in relation to business objectives or operations impacted by events, making them useful as the basis for quantitative risk assessment. Each identified risk should be included in one or more scenarios, and each scenario should be based on an identified risk.

2.4.1 Risk Scenario Development Tools and Techniques

The development of risk scenarios is based on describing a potential risk event and documenting the factors and areas that may be affected by the risk event. Each scenario should be related to a business objective or impact. Risk events include system failure, loss of key personnel, theft, network outages, power failures, natural disasters or any other situation that could affect business operations and mission. The key to developing effective scenarios is to focus on real and relevant potential risk events. Examples of this would be to develop a risk scenario based on a radical change in the market for an organization's products, a change in government or leadership or a supply chain failure.

Risk scenarios based purely on imagination require creativity, thought, consultation and questioning. Incidents that have occurred previously may be used as the basis of risk scenarios with far less effort put into their development. While risk scenarios based on past events should be fully explored to ensure that similar situations do not recur in ways that might have been avoided, sole reliance and focus cannot be devoted to only that which has already occurred but balanced to consider those scenarios which have not occurred, but have the potential likelihood of occurrence.

Risk scenarios can be developed from a top-down perspective driven by business goals or from a bottom-up perspective originating from hypothetical scenarios, as shown in **figure 2.16**.

CHAPTER 2— IT RISK ASSESSMENT

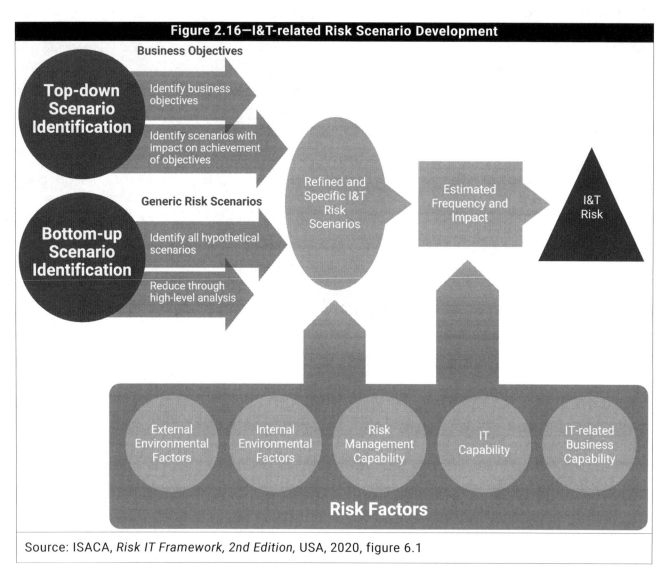

Figure 2.17 shows an example of the many inputs that are required to develop risk scenarios.

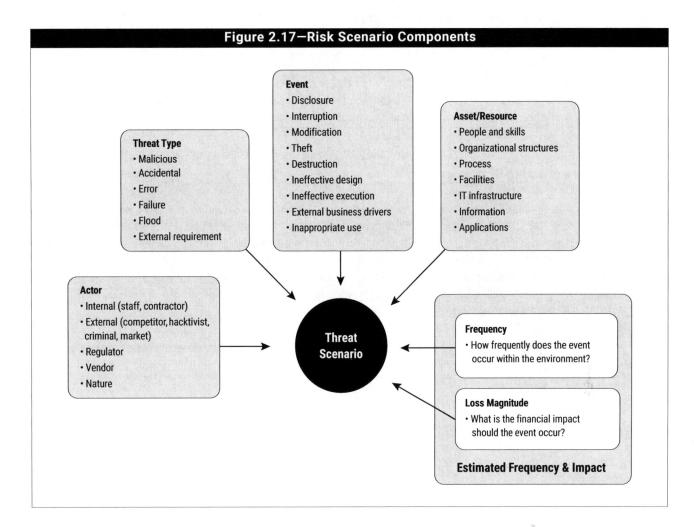

Figure 2.17—Risk Scenario Components

Top-down Approach

A top-down approach to scenario development is based on understanding business goals and how a risk event could affect the achievement of those goals. Under this model, the risk practitioner looks for the outcome of events that may hamper business goals identified by senior management. Various scenarios are developed that allow the enterprise to examine the relationship between the risk event and the business goals, so that the impact of the risk event can be measured. By directly relating a risk scenario to the business, senior managers can be educated and involved in how to understand and measure risk.

The top-down approach is suited to general risk management of the company, because it looks at both IT- and non-IT-related events. A benefit of this approach is that because it is more general, it is easier to achieve management buy-in even if management usually is not interested in IT. The top-down approach also deals with the goals that senior managers have already identified as important to them.

Bottom-up Approach

The bottom-up approach to developing risk scenarios is based on describing risk events that are specific to individual enterprise situations, typically hypothetical situations envisioned by the people performing the job functions in specific processes. The risk practitioner and assessment team start with one or more generic risk scenarios, then refine them to meet their individual organizational needs including building complex scenarios to account for coinciding events.

Bottom-up scenario development can be a good way to identify scenarios that are highly dependent on the specific technical workings of a process or system, which may not be apparent to anyone who is not intimately involved with that work but could have substantial consequences for the enterprise. One downside of bottom-up scenario development is that it may be more difficult to maintain management interest in highly specialized technical scenarios.

2.4.2 Benefits of Using Risk Scenarios

Risk scenarios facilitate communication in risk management by constructing a narrative that can inspire people to act. The use of risk scenarios can enhance the risk management effort by helping the risk team to understand and explain risk to the business process owners and other stakeholders. Additionally, a well-developed scenario provides a realistic and practical view of risk that is more aligned with business objectives, historical events and emerging threats envisioned by the enterprise than would be found by consulting a broadly applicable standard or catalog of controls. These benefits make risk scenarios valuable as means of gathering and framing information used in subsequent steps in the risk-management process.

2.4.3 Developing IT Risk Scenarios

A risk scenario is a description of an IT-related risk event that can lead to a business impact. The risk scenario contains the following components:

- Actor/threat community: The internal or external party or entity that generates the threat
- Intent/motivation: The nature of the threat event (malicious or accidental; a natural event; an equipment or process failure)
- Threat event: The security incident, such as the disclosure of information, the interruption of a system or project, including:
    - Theft
    - Improper modification of data or a process
    - Inappropriate use of resources
    - Changes to regulations
    - Lack of change management
- Asset/resource: The entity affected by the risk event, including:
    - People
    - Organizational structure
    - IT processes
    - Physical infrastructure
    - IT infrastructure
    - Information, applications
- Effect
    - How much loss would the enterprise feel as a result of a risk scenario being realized
    - What primary losses could be experienced if a risk scenario were realized
    - What secondary losses could be experienced if a risk scenario were realized
- Timing:

CHAPTER 2— IT RISK ASSESSMENT

- How often the threat agent and asset are in contact
- How often a threat agent acts against an asset
- How often the threat agent is successful in overcoming existing controls
- How often the controls are successful in preventing a successful action
- How often will a loss be realized

Some key issues related to the development of risk scenarios are addressed in **figure 2.18**.

Figure 2.18—Risk Scenario Technique Main Focus Areas	
Focus/Issue	Summary Guidance
Maintain currency of risk scenarios and risk factors.	Risk factors and the enterprise change over time; hence, scenarios will change over time, over the course of a project or over the evolution of technology.
	For example, it is essential that the risk function develop a review schedule and the chief information officer works with the business lines to review and update scenarios for relevance and importance. Frequency of this exercise depends on the overall risk profile of the enterprise and should be done at least on an annual basis, or when important changes occur.
Use generic scenarios as a starting point and build more detail where and when required.	One technique of keeping the number of scenarios manageable is to propagate a standard set of generic scenarios throughout the enterprise and develop more detailed and relevant scenarios when required and warranted by the risk profile only at lower (entity) levels. The assumptions made when grouping or generalizing should be well understood by all and adequately documented because they may hide certain scenarios or be confusing when looking at risk response.
	For example, if "insider threat" is not well-defined within a scenario, it may not be clear whether this threat includes privileged and non-privileged insiders. The differences between these aspects of a scenario can be critical when one is trying to understand the frequency and impact of events and mitigation opportunities.
Number of scenarios should be representative and reflect business reality and complexity.	Risk management helps to deal with the enormous complexity of IT environments by prioritizing potential action according to its value in reducing risk. Risk management is about reducing complexity, not generating it; hence, another plea for working with a manageable number of risk scenarios. However, the retained number of scenarios still needs to accurately reflect business reality and complexity.
Risk taxonomy should reflect business reality and complexity.	There should be a sufficient number of risk scenario scales reflecting the complexity of the enterprise and the extent of exposures to which the enterprise is subject.
	Potential scales should be aligned and communicated throughout the enterprise to ensure consistent use.
Use generic risk scenario structure to simplify risk reporting.	Similarly, for risk reporting purposes, entities should not report on all specific and detailed scenarios but could do so by using the generic risk structure.
	For example, an entity may have taken generic scenario 15 (project quality), translated it into five scenarios for its major projects, subsequently conducted a risk analysis for each of the scenarios then aggregated or summarized the results and reported back using the generic scenario header "project quality."

CHAPTER 2— IT RISK ASSESSMENT

Figure 2.18—Risk Scenario Technique Main Focus Areas *(cont.)*	
Focus/Issue	Summary Guidance
Ensure adequate people and skills requirements for developing relevant risk scenarios.	Developing a manageable and relevant set of risk scenarios requires: • Expertise and experience to not overlook relevant scenarios and not be drawn into highly unrealistic or irrelevant scenarios. While the avoidance of scenarios that are unrealistic or irrelevant is important in properly utilizing limited resources, some attention should be paid to situations that are highly infrequent and unpredictable, but which could have a cataclysmic impact on the enterprise. • A thorough understanding of the environment. This includes the IT environment (e.g., infrastructure, applications, dependencies between applications, infrastructure components), the overall business environment and an understanding of how and which IT environments support the business environment to understand the business impact. • The intervention and common views of all parties involved—senior management, which has the decision power; business management, which has the best view on business impact; IT, which understands what can go wrong with IT; and risk management, which can moderate and structure the debate amongst the other parties. • The process of developing scenarios usually benefits from a brainstorming/workshop approach, where a high-level assessment is required to reduce the number of scenarios to a manageable, but relevant and representative number.
Use the risk scenario building process to obtain buy-in.	Scenario analysis is not just an analytical exercise involving "risk analysts." A significant additional benefit of scenario analysis is achieving organizational buy-in from enterprise entities and business lines, risk management, IT, finance, compliance and other parties. Gaining this buy-in is the reason why scenario analysis should be a carefully facilitated process.
Involve first line of defense in the scenario building process.	In addition to coordinating with management, it is recommended that selected members of the staff who are familiar with the detailed operations be included in discussions, where appropriate. Staff whose daily work is in the detailed operations are often more familiar with vulnerabilities in technology and processes that can be exploited.
Do not focus only on rare and extreme scenarios.	When developing scenarios, one should not focus only on worst-case events because they rarely materialize, whereas less-severe incidents happen more often.
Deduce complex scenarios from simple scenarios by showing impact and dependencies.	Simple scenarios, once developed, should be fine-tuned into more complex scenarios, showing cascading and/or coincidental impacts and reflecting dependencies. For example: • A scenario of having a major hardware failure can be combined with the scenario of failed disaster recovery plan. • A scenario of major software failure can trigger database corruption and, in combination with poor data management backups, can lead to serious consequences, or at least consequences of a different magnitude than a software failure alone. • A scenario of a major external event can lead to a scenario of internal apathy.
Consider systemic and contagious risk.	Attention should be paid to systemic and/or contagious risk scenarios: • **Systemic**—Something happens with an important business partner, affecting a large group of enterprises within an area or industry. An example would be a nationwide air traffic control system that goes down for an extended period of time (e.g., six hours) affecting air traffic on a very large scale. • **Contagious**—Events that happen at several of the enterprise's business partners within a very short time frame. An example would be a clearinghouse that can be fully prepared for any sort of emergency by having very sophisticated disaster recovery measures in place, but when a catastrophe happens, finds that no transactions are sent by its providers and is temporarily out of business.

CHAPTER 2— IT RISK ASSESSMENT

Figure 2.18—Risk Scenario Technique Main Focus Areas (cont.)	
Focus/Issue	Summary Guidance
Use scenario building to increase awareness for risk detection.	Scenario development also helps to address the issue of detectability, moving away from a situation where an enterprise "does not know what it does not know." The collaborative approach for scenario development assists in identifying risk to which the enterprise, until then, would not have realized it was subject to (and hence would never have thought of putting in place any countermeasures). After the full set of risk items is identified during scenario generation, risk analysis assesses frequency and impact of the scenarios. Questions to ask include: • Will the enterprise ever detect that the risk scenario has materialized? • Will the enterprise notice something has gone wrong so it can react appropriately? Generating scenarios and creatively thinking of what can go wrong will automatically raise and, hopefully, cause response to the question of detectability. Detectability of scenarios includes two steps: visibility and recognition. The enterprise must be in a position that it can observe anything going wrong, and it needs the capability to recognize an observed event as something wrong.
Source: ISACA, *COBIT 5 for Risk*, USA, 2013, figure 37	

Business continuity teams are often working from similar types of scenarios as those developed for purposes of risk identification and assessment, and the risk practitioner may find it beneficial to partner with these teams as a means of avoiding unnecessary duplication of effort. Reviewing audit reports, interviewing end users or IT personnel, and observing the business processes or operations in action may also provide a useful basis for scenario development.

2.4.4 Analyzing Risk Scenarios

During risk identification, risk scenarios are developed and used to identify and describe potential risk events. These scenarios are useful to communicate with the business and gather input data required to understand the potential or probable impact of the risk event if it were to occur.

The impact of a risk event is hard to calculate with any degree of precision because there are many factors that affect the outcome of an event. If the event is detected quickly and appropriate measures are taken to contain the incident, then the impact may be minimized, and the recovery process may be fairly rapid. However, if the enterprise is unable to detect the incident promptly, the same incident could cause severe damage and result in much higher recovery costs.

The Factor Analysis of Information Risk (FAIR) Model (**Figure 2.19**), which decomposes the major components that comprise risk into smaller, manageable components ready to be analyzed.

CHAPTER 2— IT RISK ASSESSMENT

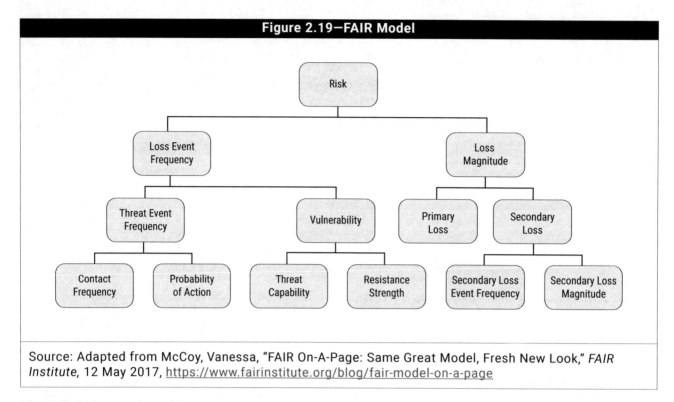

Figure 2.19—FAIR Model

Source: Adapted from McCoy, Vanessa, "FAIR On-A-Page: Same Great Model, Fresh New Look," *FAIR Institute*, 12 May 2017, https://www.fairinstitute.org/blog/fair-model-on-a-page

The Holistic Approach to Risk Management (HARM) is an entire methodology that is built on the OpenFAIR model and is similar in nature, but accounts for loss magnitudes at a discrete level and factors in control objective maturity as a method to account for potential reductions in overall loss magnitude estimations (**figure 2.20**).

CHAPTER 2— IT RISK ASSESSMENT

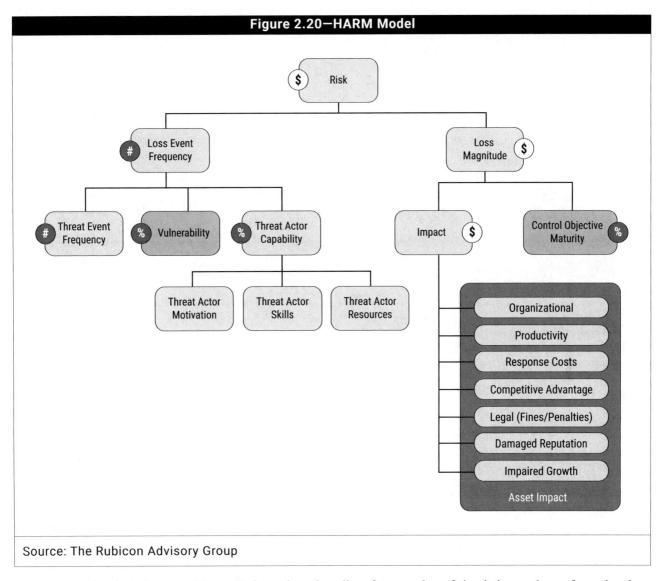

Figure 2.20—HARM Model

Source: The Rubicon Advisory Group

Both FAIR and HARM leverage Monte Carlo engines that allow for a number of simulations to be performed and map qualitative statements to quantitative values.

Page intentionally left blank

Part B: IT Risk Analysis, Evaluation and Assessment

After risk is identified and documented in the risk register, the next step in the risk management life cycle, shown in **figure 2.21**, is to assess the IT risk level. IT risk is a subset of enterprise risk, and the risk faced by an IT system is most often measured by the impact of an IT-related problem on the business services that the IT system supports. Calculations or assessments of IT impact must consider the dependencies of the other systems, departments, business partners and users on the affected IT system.

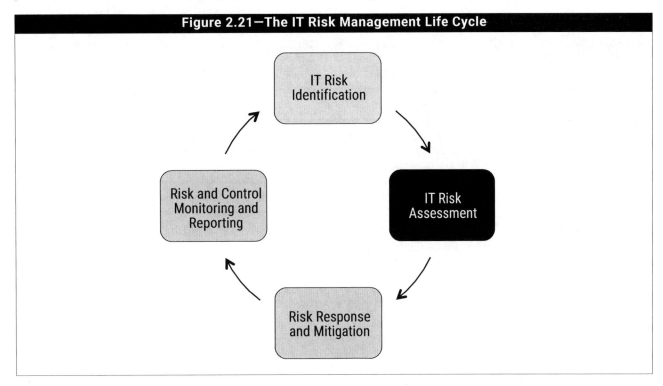

Figure 2.21—The IT Risk Management Life Cycle

Risk assessment is a process used to identify and evaluate risk and its potential effects, which includes evaluation of the:

- Critical functions necessary for an enterprise to continue business operations
- Risk associated with each of the critical functions
- Controls in place to reduce exposure and their cost
- Prioritization of the risk on the basis of their likelihood and potential impact
- Relationship between the risk and the enterprise risk appetite and tolerance

The risk assessment process generates the information used to respond to risk in an appropriate and cost-effective manner.

2.5 Risk Assessment Concepts, Standards and Frameworks

There are several techniques used in risk assessment, including those listed in **figure 2.22**.

CHAPTER 2— IT RISK ASSESSMENT

Figure 2.22—Risk Assessment Techniques	
Technique	Description
Bayesian analysis	A Bayesian analysis is a method of statistical inference that uses prior distribution data to determine the probability of a result. This technique relies on the prior distribution data to be accurate in order to be effective and to produce accurate results.
Bow tie analysis	A bow tie analysis provides a diagram to communicate risk assessment results by displaying links between possible causes, controls and consequences. The cause of the event is depicted in the middle of the diagram (the "knot" of the bow tie) and triggers, controls, mitigation strategies and consequences branch off of the "knot."
Brainstorming/structured interview	The structured interview and brainstorming model gathers a large group of types of potential risk or ideas to be ranked by a team. The initial interview or brainstorming may be completed using prompts or interviews with an individual or small group.
Cause and consequence analysis	A cause and consequence analysis combines techniques of a fault tree analysis and an event tree analysis and allows for time delays to be considered.
Cause-and-effect analysis	A cause-and-effect analysis looks at the factors that contributed to a certain effect and groups the causes into categories (using brainstorming), which are then displayed using a diagram, typically a tree structure or a fishbone diagram.
Checklists	A checklist is a list of potential or typical threats or other considerations that should be of interest to the organization, whose items can be checked off one at a time as they are completed. The risk practitioner may use previously developed lists, codes or standards to assess the risk using this method.
Delphi method	The Delphi method leverages expert opinion received using two or more rounds of questionnaires. After each round of questioning, the results are summarized and communicated to the experts by a facilitator. This collaborative technique is often used to build a consensus among experts.
Event tree analysis	An event tree analysis is a forward-looking, bottom-up model that uses inductive reasoning to assess the probability of different events resulting in possible outcomes.
Factor Analysis of Information Risk (FAIR)	FAIR is a taxonomy of the factors that contribute to risk and how they affect each other. It is primarily concerned with establishing accurate probabilities for the frequency and magnitude of data loss events.
Fault tree analysis	A fault tree analysis starts with an event and examines possible means for the event to occur (top-down) and displays these results in a logical tree diagram. This diagram can be used to generate ways to reduce or eliminate potential causes of the event.
Hazard analysis and critical control points (HACCP)	Originally developed for the food safety industry, HACCP is a system for proactively preventing risk and assuring quality, reliability and safety of processes. The system monitors specific characteristics, which should fall within defined limits.
Hazard and operability studies (HAZOP)	HAZOP is a structured means of identifying and evaluating potential risk by looking at possible deviations from existing processes.
Holistic Approach to Risk Management (HARM)	Based on the OpenFAIR model, HARM uses the FAIR taxonomy, using threat scenarios but taking a process-based approach that allows for an OpenFAIR analysis to be performed, relying on the ability to demonstrate control implementation and control objective maturity.
Human reliability analysis (HRA)	HRA examines the effect of human error on systems and their performance.
Layers of protection analysis (LOPA)	LOPA is a semi-quantitative risk analysis technique that uses aspects of HAZOP data to determine risk associated with risk events. It also looks at controls and their effectiveness.
The lotus blossom method of brainstorming	Lotus blossom is a brainstorming technique that makes use of visual representations by adding the problem in the center of the "blossom" of a 3×3 square. Then, related solutions or themes are added in an iterative manner, expanding outwards.
Markov analysis	A Markov analysis is used to analyze systems that can exist in multiple states. The Markov model assumes that future events are independent of past events.

CHAPTER 2— IT RISK ASSESSMENT

Figure 2.22—Risk Assessment Techniques *(cont.)*	
Technique	Description
Monte-Carlo analysis	IEC 31010:2009 describes Monte Carlo simulation in the following manner: *Monte Carlo simulation is used to establish the aggregate variation in a system resulting from variations in the system, for a number of inputs, where each input has a defined distribution and the inputs are related to the output via defined relationships. The analysis can be used for a specific model where the interactions of the various inputs can be mathematically defined. The inputs can be based upon a variety of distribution types according to the nature of the uncertainty they are intended to represent. For risk assessment, triangular distributions or beta distributions are commonly used.*
OCTAVE (Operationally Critical Threat, Asset and Vulnerability Evaluation)	OCTAVE is a flexible and self-directed risk assessment methodology. A small team of people from the operational (or business) units and the IT department work together to address the security needs of the enterprise. The team draws on the knowledge of many employees to define the current state of security, identify risk to critical assets and set a security strategy. It can be tailored for most enterprises.
Preliminary hazard analysis	Preliminary hazard analysis looks at what threats or hazards may harm an organization's activities, facilities or systems. The result is a list of potential risk.
Reliability-centered maintenance	Reliability-centered maintenance analyzes the functions and potential failures of a specific asset, particularly a physical asset such as equipment.
Sneak circuit analysis	A sneak circuit analysis is used to identify design errors or sneak conditions such as latent hardware, software or integrated conditions that are often undetected by system tests and may result in improper operations, loss of availability, program delays or injury to personnel.
Structured "what if" technique (SWIFT)	A structured "what if" technique uses structured brainstorming to identify risk, typically within a facilitated workshop. It uses prompts and guide words and is typically used with another risk analysis and evaluation technique.

A consistent risk assessment technique should be used whenever the goal is to produce results that can be compared over time. However, risk practitioners need not rely exclusively on one (or even several) techniques. Each approach has certain advantages and possible weaknesses, and the risk practitioner should choose a technique appropriate for the circumstances of the assessment.

2.5.1 Risk Ranking

The risk practitioner uses the results of risk assessment to place risk in an order that can be used to direct the risk response effort. Risk ranking is derived from a combination of all the components of risk including the recognition of the threats and the characteristics and capabilities of a threat source, the severity of a vulnerability, the likelihood of attack success when considering effectiveness of controls and the impact to the organization of a successful attack. Taken together, these indicate the level of risk associated with a threat.

Figure 2.23 shows an example of a matrix used to document risk rankings.

CHAPTER 2— IT RISK ASSESSMENT

Figure 2.23—Template—Adversarial Risk												
1	2	3	4	5	6	7	8	9	10	11	12	13
Threat Event	Threat Sources	Threat Source Characteristics			Relevance	Likelihood of Attack Initiation	Vulnerabilities and Predisposing Conditions	Severity and Pervasiveness	Likelihood Initiated Attack Succeeds	Overall Likelihood	Level of Impact	Risk
		Capability	Intent	Targeting								

Source: NIST, *NIST Special Publication 800-30 Revision 1: Guide for Conducting Risk Assessments*, USA, 2012. Reprinted courtesy of the National Institute of Standards and Technology, U.S. Department of Commerce. Not copyrightable in the United States.

Risk Maps

After risk has been determined using quantitative, qualitative or semiquantitative methods, the risk practitioner needs to review whether the risk levels are within the boundaries of acceptable risk as defined by senior management through the setting of the organizational risk appetite and risk tolerance.

Where risk analysis is the process of determining probability and impact of a given scenario, risk evaluation is the process to consider that scenario in context of the organization's defined risk appetite and criteria used for risk tolerance. It is during this process where, understanding the organization's definition of acceptable and unacceptable risk, that we evaluate the event in the context of the organization risk appetite, tolerance and capacity.

This is commonly visualized and represented by mapping each threat, in context of likelihood and impact, from acceptable through unacceptable, with probability and impact depicted on the axis to provide a frame of reference into a risk map (**figure 2.24**).

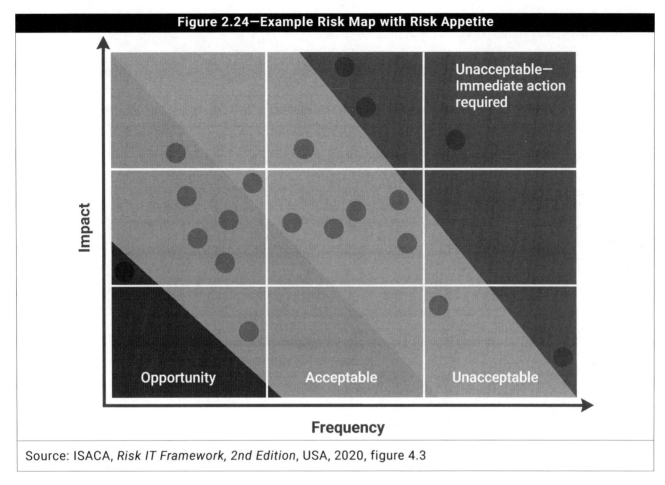

Figure 2.24—Example Risk Map with Risk Appetite

Source: ISACA, *Risk IT Framework, 2nd Edition*, USA, 2020, figure 4.3

Based upon where a scenario lies within the risk map, the practitioner can then make an assessment as to what the appropriate risk response should be. In this illustration we see the organization has defined four layers ranging from Opportunity (continues improvement and reduction of risk severity) to Unacceptable—Immediate Action Required (consider this the risk capacity) along with providing details of Frequency (x-axis) and Impact (y-axis).

Understanding where a scenario falls on the risk map can quickly focus efforts to the risk that falls outside of the acceptable risk boundaries and prevents spending an undue amount of time on the risk that is within defined, acceptable levels.

The risk assessment process is used to determine the appropriate response to address risk, given the probability and potential effects in context of the enterprise's risk appetite, tolerance, and capacity, which consider:

- Efficiency of the response
- Exposure of the risk at present
- Implementation capability
- Effectiveness of the response

2.5.2 Risk Ownership and Accountability

Each risk must be linked to an individual who owns the risk, according to their job responsibilities and duties. The risk owner is tasked with and responsible for making the decision of what the best response is to the identified risk and must be at a level in the enterprise where they are authorized to make decisions on behalf of the enterprise and can be held accountable for those decisions. To ensure accountability, the ownership of risk must be with an individual, not with a department or the enterprise as a whole.

2.5.3 Documenting Risk Assessments

At the conclusion of the IT risk assessment phase, the risk practitioner compiles the results of the risk assessment into a comprehensive report for senior management. The risk assessment report should indicate any gaps between the current risk environment and the desired state of IT risk, advise whether these gaps are within acceptable levels, and provide some basis on which to judge the severity of the identified issue. The risk assessment report should document the process used as well as the result of the risk assessment. Each risk should be documented in a manner that is understandable to management and clearly states the risk levels and priorities.

The risk assessment report may be adjusted according to the needs of the enterprise and the direction of senior management. A typical risk assessment report includes the following sections:

- Objectives of the risk assessment process
- Scope and description of the area subject to assessment
- External context and factors affecting risk
- Internal factors or limitations affecting risk assessment
- Risk assessment criteria
- Risk assessment methodology used
- Resources and references used
- Identification of risk, threats and vulnerabilities
- Assumptions used in the risk assessment
- Potential of unknown factors affecting assessment
- Results of risk assessment
- Recommendations and conclusions

The risk assessment should be performed in a consistent manner which supports future risk assessment efforts and would provide predictable results. All risk should be noted in the report, including issues that may already have been addressed, in order to provide an accurate picture of risk to senior management. This comprehensive approach ensures that if a control that was in place at the time of the assessment is later removed, the risk practitioner is not suspected of overlooking it during the identification and assessment processes.

Enterprises assess IT risk because IT risk is a form of business risk, not because it is a nuisance to the IT staff. The risk practitioner should refrain from using terminology that is specific to IT or may be misinterpreted by management. Many terms and acronyms commonly used in IT have different meaning to other disciplines in the enterprise, which may result in the misunderstanding of the problem and the proposed solutions. The risk practitioner should ensure that the report is clear, concise and accurate and is free from terminology that could be misunderstood or is subject to misinterpretation.

2.5.4 Addressing Risk Exclusions

The risk practitioner should take care to ensure that all IT risk is either evaluated or intentionally excluded (according to the scope of the risk assessment and the relevance of the risk to organizational goals or assets). Some IT risk events apply only to enterprises that meet particular criteria. For instance, not all regions are subject to the same natural disasters. Likewise, the risk of losing credit card data applies only to those enterprises that store or process credit card data. When an assumption is made to not include a particular risk in an assessment, the risk

CHAPTER 2— IT RISK ASSESSMENT

practitioner may find it beneficial to document that the risk was intentionally excluded rather than missed in the risk assessment process and provide justification and rationale behind the exclusion.

In general, this occurs during the risk identification process, but as a deeper level of investigation and evaluation of risk is conducted, new types of risk may surface while others that were previously excluded are found to be relevant. For this reason, the risk practitioner should reevaluate each documented risk to ensure that it was identified and assessed accurately based on the current risk landscape.

2.6 Risk Register

The purpose of a risk register is to consolidate risk data into one place and enable the tracking of risk. Entries in the risk register show the severity, source and potential impact of a risk and identify the risk owner and the current status and disposition of the risk. The risk register is the one document that contains all risk that has been detected by various parts and activities of the enterprise including risk identified in audits, vulnerability assessments, penetration tests, incident reports, process reviews, management input, risk scenario creation and security assessments. This allows management to refer to one place to gain insight into the outstanding risk issues, the status of risk mitigation efforts and the emergence of newly identified and documented risk. In some implementations, the risk register may be accompanied by a controls register, which includes information on testing and monitoring.

Enterprises maintain risk registers in order to consolidate and track relevant information about risk into one central repository. The risk register allows senior management and the managers of each department to obtain the status of the risk-management process from a single source, which in turn makes it possible to better manage and report on risk and coordinate risk response activities. At any given time, the risk register should document the entire risk universe of the enterprise.

The risk register must be kept up to date in order to be effective, benefit and value. Risk practitioners should be familiar with the fundamental elements of a risk register and ensure that its content is periodically reviewed both to add new risk and re-assess risk to ensure relevancy to the enterprise. An example of a risk register template is shown in **Figure 2.25**.

Figure 2.25—Example of a Risk Register Template				
Part I—Summary Data				
Risk statement				
Risk owner				
Date of last risk assessment				
Due date for update of risk assessment				
Risk category	Strategic	Program	Project	Operational
Risk classification (copied from risk analysis results)	☐ LOW	☐ MEDIUM	☐ HIGH	☐ VERY HIGH
Risk response	☐ ACCEPT	☐ TRANSFER	☐ MITIGATE	☐ AVOID
Part II—Risk Description				
Title				

CHAPTER 2— IT RISK ASSESSMENT

Figure 2.25—Example of a Risk Register Template (cont.)						
High-level scenario (from list of sample high-level scenarios)						
Detailed scenario description— Scenario components	Actor					
	Threat Type					
	Event					
	Asset/Resource					
	Probability					
	Consequences					
Other scenario information						
Part III—Risk Analysis Results							
Frequency of scenario (number of times per year)	0	1	2	3	4	5
	N ≤ 0,01 ☐	0,01 < N ≤ 0,1 ☐	0,1 < N ≤ 1 ☐	1 < N ≤ 10 ☐	10 < N ≤ 100 ☐	100 < N ☐
Comments on frequency						
Impact of scenario on business	0	1	2	3	4	5
1. Productivity	Revenue Loss Over One Year						
Impact rating	I ≤ 0,1% ☐	0,1% < I ≤ 1% ☐	1% < I ≤ 3% ☐	3% < I ≤ 5% ☐	5% < I ≤ 10% ☐	10% < I ☐
Detailed description of impact						
2. Cost of response	Expenses Associated With Managing the Loss Event						
Impact rating	I ≤ 10k$ ☐	10K$ < I ≤ 100K$ ☐	100K$ < I ≤ 1M$ ☐	1M$% < I ≤ 10M$ ☐	10M$ < I ≤ 100M$ ☐	100M$ < I ☐
Detailed description of impact						
3. Competitive advantage	Drop-in Customer Satisfaction Ratings						
Impact rating	I ≤ 0,5 ☐	0,5 < I ≤ 1 ☐	1 < I ≤ 1,5 ☐	1,5 < I ≤ 2 ☐	2 < I ≤ 2,5 ☐	2,5 < I ☐
Detailed description of impact						
4. Legal	Regulatory Compliance—Fines						
Impact rating	None ☐	< 1M$ ☐	< 10M$ ☐	< 100M$ ☐	< 1B$ ☐	> 1B$ ☐
Detailed description of impact						
Overall impact rating (average of four impact ratings)						
Overall rating of risk (obtained by combining frequency and impact ratings on risk map)	☐ LOW	☐ MEDIUM		☐ HIGH	☐ VERY HIGH	
Part III—Risk Response							
Risk response for this risk	☐ ACCEPT	☐ TRANSFER		☐ MITIGATE	☐ AVOID	
Justification						

Figure 2.25—Example of a Risk Register Template (cont.)

Detailed description of response (NOT in case of ACCEPT)	Response Action	Completed	Action Plan
	1.	☐	☐
	2.	☐	☐
	3.	☐	☐
	4.	☐	☐
	5.	☐	☐
	6.	☐	☐
Overall status of risk action plan			
Major issues with risk action plan			
Overall status of completed responses			
Major issues with completed responses			
Part IV—Risk Indicators			
Key risk indicators for this risk	1.		
	2.		
	3.		
	4.		
Source: ISACA, *COBIT 5 for Risk*, USA, 2013, figure 62			

2.7 Risk Analysis Methodologies

Risk analysis is a complex and important process that is often needed to provide the data necessary for effective risk response. Risk may be analyzed on either a quantitative or qualitative basis or in a way that reflects some combination of the two (semiquantitative/hybrid model).

2.7.1 Quantitative Risk Assessment

Quantitative risk assessment leverage scenarios which tend to leverage numerical calculations and use common mathematical models to simulate potential outcomes and are often represented in monetary values. Its reliance on quality data tend to make it more accurate when performing calculations leading to a level of precision. It is particularly suitable for cost-benefit analysis because risk that can be mapped to monetary values can easily and directly be compared to the costs of various risk responses. The challenges of calculating likelihood requires the risk practitioner become familiar with the variables that should be considered.

Where quantitative risk assessment is desirable, the risk practitioner may seek to approximate probability using calibrated estimates in addition to quality empirical or historical data to model, simulate and calculate a likelihood over an entire population. For instance, although MTBF is not an accurate predictor of failure for an individual appliance, an organization that has thousands of identical appliances in use can reasonably expect some number of them to begin failing at or around the time indicated by the MTBF for their make and model.

Quantitative risk assessment becomes progressively more useful as likelihood approaches certainty, which is a scenario that can be applied to hurricanes, blizzards and other sorts of weather events that are prevalent in certain geographical regions. In such cases, the need to account for the probability of an event is de-emphasized, because the event itself may be assumed. The risk practitioner is then left to account only for the impact to the organization of the associated consequences, which may be measured very precisely in terms of direct cost associated with damage to property and loss of revenue.

The value of risk used in quantitative risk assessment is often calculated on an annual basis in order to align the process with the natural cycle for calculating budgets. The risk practitioner may find it beneficial to compare various events that occur at different frequencies in a similar manner using a common denominator.

2.7.2 Qualitative Risk Assessment

Qualitative risk assessment assigns values on a comparative or ordinal basis (such as high, medium and low or a scale of 1 to 10). The assignment of qualitative values relies heavily on experience and expert knowledge, resulting in subjective findings. Many types of risk may appear to be extremely difficult to calculate in purely numeric terms, and based on the maturity of the organization and experience of the risk practitioner, a quantitative assessment may be difficult to conduct. In contrast, the relative values offered by a qualitative process can typically be used to order response actions in terms of perceived importance. When the perception is based on a broad enough sample of stakeholders, the resulting course of action is likely to be acceptable to those stakeholders.

Qualitative risk assessments are usually based on scenarios or descriptions of situations that either have occurred or may occur. The intention of these scenarios is to elicit feedback from multiple stakeholders (e.g., departments, customers, management). By communicating with all affected shareholders, the risk practitioner is able to eliminate unlikely assumptions and determine a level of risk that reflects a reasonable consensus of feedback from all potentially affected groups.

Scenarios under qualitative risk assessment may be based on threats, vulnerabilities, assets/impact or some combination of these factors. A threat-based scenario examines a risk event on the basis of threat actors and seeks to identify potential methods of attack. From this information, it can be inferred which vulnerabilities will be exploited by the attacker along with the intent and skill of the attacker and the potential damage to the assets affected. This method is especially beneficial when examining the emergence of new threats and determining the risk related to APTs, whose capabilities and available time make waiting for suitable vulnerabilities more feasible than is true of conventional attackers.

A vulnerability-based approach takes the opposite approach: it examines the organization's known vulnerabilities and then attempts to anticipate threats that could exploit those vulnerabilities, projecting from these the consequences and magnitude of impact. Vulnerability-based scenarios are especially valuable after completing vulnerability assessments, and their findings can be further assessed by carrying out penetration testing to determine whether the anticipated threats are feasible in the context of the organization and its systems.

An asset/impact approach is based on the identification of critical and sensitive assets and the potential ways that these could be damaged—typically by attacking their confidentiality, integrity and availability.

The results of a qualitative risk assessment are typically presented in the form of a table that compares the likelihood of a risk event with its impact to the enterprise, where the confluence of the two factors generates the relative level of the risk. For instance, a risk that is highly likely and has a high level of impact would be identified as an area of immediate concern, whereas a risk of lower likelihood or a lower level of exposure would represent a lower level of priority for risk response.

2.7.3 Semiquantitative/Hybrid Risk Assessment

Semiquantitative risk assessment combines the value of qualitative and quantitative risk assessment. A hybrid approach has the realistic input of a qualitative assessment combined with the numerical scale used to determine the impact of a quantitative risk assessment. The goal is to provide a scale with a wide enough range that risk can be assessed in a fairly precise manner, such as 0 to 100. This numeric values make it easier to prioritize a large number of risk factors than is true with a typical qualitative range of three to five levels. However, the risk practitioner should take care to ensure that risk owners do not mistake the origins of these values as coming from purely objective sources.

An example of semiquantitative risk assessment is shown in **figure 2.26**.

Figure 2.26—Assessment Scale—Level of Risk			
Qualitative Values	Semi-quantitative Values		Description
Very High	96-100	10	Very high risk means that a threat event could be expected to have multiple severe or catastrophic adverse effects on organizational operations, organizational assets, individuals, other organizations or the nation.
High	80-95	8	High risk means that a threat event could be expected to have a severe or catastrophic adverse effect on organizational operation, organizational assets, individuals, other organizations or the nation.
Moderate	21-79	5	Moderate risk means that a threat event could be expected to have a serious adverse effect on organizational operations, organizational assets, individuals or the nation.
Low	5-20	2	Low risk means that a threat event could be expected to have a limited adverse effect on organizational operations, organizational assets, individuals, other organizations or the nation.
Very Low	0-4	0	Very low risk means that a threat event could be expected to have a negligible adverse effect on organizational operations, organizational assets, individuals, other organizations or the nation.
Source: National Institute of Standards and Technology (NIST), *NIST Special Publication 800-30 Revision 1: Guide for Conducting Risk Assessments*, USA, 2012. Reprinted courtesy of the National Institute of Standards and Technology, U.S. Department of Commerce. Not copyrightable in the United States.			

Semi quantitative risk assessment can be an effective compromise when impact is quantifiable, but likelihood is not. Under such circumstances, applying a basic range of high, medium and low may not offer sufficient precision to generate useful risk ratings, whereas a more granular range of likelihood values may be used with the quantified impact to support specific recommendations for risk response.

2.8 Business Impact Analysis

Business impact analysis (BIA) is a process to establish continuity requirements and determine the impact of losing the support of any resource. In addition to identifying initial impact, a comprehensive BIA seeks to establish the escalation of loss over time. The goal of BIA is to provide reliable data on the basis of which senior management can make the appropriate decision.

Why are BIAs so important? When considering business continuity as another management aspect of organizational resilience, BIAs are the process in which requirements for the continuation of critical business services are defined and prioritized. Just as a risk practitioner would not start an assessment without clearly identifying the component parts of a threat scenario, the same is true for business continuity; BIAs need to define clear requirements.

Outcomes associated with completing a BIA include:
- The ability to prioritize business defined critical services, which support strategic priorities, goals and objectives
- The ability to determine how business critical services should be protected

CHAPTER 2— IT RISK ASSESSMENT

- The ability to define the acceptable levels of diminished operation before impacting strategic priorities
- The ability of the organization to achieve its goal and objectives.

These abilities can then be transformed into defining:

- The business activities and resources required to achieve the enterprise's strategy
- What needs to be protected and order of recovery following an incident
- Recovery timeframes that in turn can inform when services need to be recovered

The BIAs conducted by organization can aid the risk practitioner in recommending reasonable and appropriate risk response and guide senior management in selecting appropriate treatment and recovery strategies.

Additionally, recovery time objectives (RTOs) should be established in such a way that, if achieved, would enable an enterprise to meet its strategic priorities. See Chapter 4 Information Technology and Security for more information.

2.8.1 Business Continuity and Organizational Resiliency

The BIA classifies business activities and resources needed for the delivery of an organization's most essential services. By understanding how an enterprise delivers services, the process should document key processes, practices, activities and resources needed to achieve the organization's goals and objectives; this also has the potential to uncover previously unknown factors that can now be accounted for and considered.

Enterprises are often compromised of a variety of separate business functions, each focused on delivering a specific service back to the organization. Sales, IT, marketing, human resources and finance are some common examples. As each business function services a specific purpose, so does its corresponding business processes. Each business function may have anywhere from a few supporting business processes to several thousand, depending on the size, maturity and complexity of the organization. **Figure 2.27** illustrates this process.

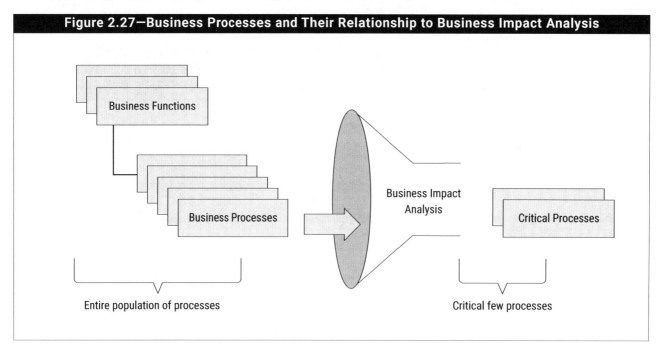

Distilling the volume of business processes to the few critical business processes that are absolutely required by the enterprise to deliver on obligations is one of the key results generated out of the BIA process.

CHAPTER 2— IT RISK ASSESSMENT

It is important to understand the various processes, practices, activities and resources (staff and technical) that may be impacted by an incident. The BIA process allows the organization to identify which key processes and resources are deemed critical, regardless of incident type, which may prevent the enterprise from successfully achieving their target goals and objectives.

2.8.2 Regulatory and Contractual Obligations

Most enterprises are often compelled, due to statutory, regulatory or contractual requirements, to ensure and maintain a certain level of availability of specific services, in light of adversity. The BIA process can be used to identify any potential gaps with those requirements and be used to build a roadmap to shore up any discrepancies. This will ensure that senior management has sufficient details around capabilities and a thorough understanding of the obligations they are required to meet. This will then enable the appropriate level of planning and preparation prior to an incident, while at the same time ensuring conformance with the requirements the organization has agreed to satisfy.

2.8.3 Strategic Investments

Understanding how, when and where to invest is always met with a healthy dose of skepticism and reviewed by many people. The ability to tie the costs of downtime is one of the many values that the BIA process yields for senior management. The deliverables generated by a BIA can foster and promote an understanding around the organizational, reputational and brand impacts not just the financial implications associated with nonconformance with contractual or regulatory obligations.

Once the wide range of impacts associated with an incident are understood, the business can develop the business cases, providing the justification for investments. The enterprise can then identify and implement the appropriate capabilities to meet recovery objectives in a financially sound manner.

2.8.4 Beyond Business Impact

The focus and purpose of the BIA process is to ultimately determine the impact that would be felt should an incident occur which disrupts normal operations.

The BIA process should also be used to inform the business continuity/organizational resiliency planning efforts, to ensure that the appropriate controls needed to safeguard those business functions and their supporting resources.

Additionally, the BIA should serve as a starting point to guided IT disaster recovery planning activities, as well as verify RTO, recovery point objectives (RPO), service delivery objectives (SDO) and maximum tolerable downtime (MTD). See Section 4.4.1 Business Continuity for more information.

Figure 2.28 further explains the relationship among these elements.

CHAPTER 2— IT RISK ASSESSMENT

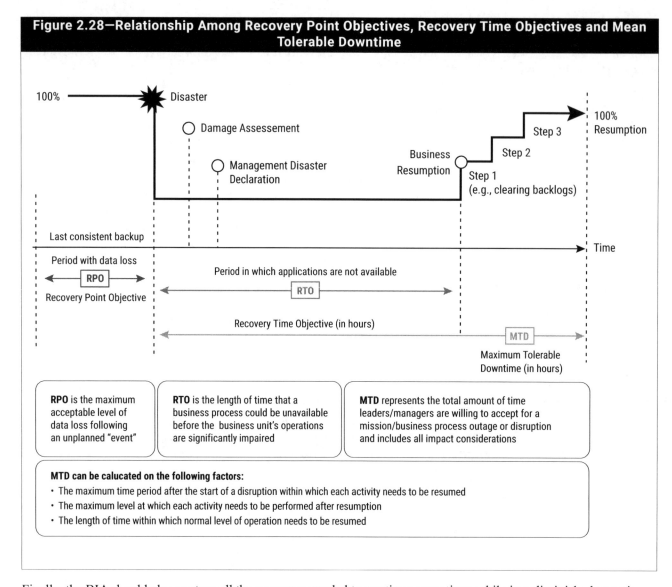

Figure 2.28—Relationship Among Recovery Point Objectives, Recovery Time Objectives and Mean Tolerable Downtime

Finally, the BIA should also capture all the resources needed to continue operations while in a diminished capacity.

While the BIA process is largely defined as a data collection effort to support and allow for analysis efforts to be performed, additional benefits can also be realized. While performing a BIA, additional information that is provided which may be of use to a risk practitioner, includes:

- Identify existing controls
- Identify current weaknesses and vulnerabilities
- Identify key threats to the business
- Document current recovery methods
- Find team and staffing requirements
- Validate and verify internal and external contact information
- Defined recovery objectives
- Identification of gaps (capabilities/systems)

2.8.5 Business Impact Analysis and Risk Assessment

BIA and risk assessment are often talked about at the same time, and that is because many enterprises perform these two activities in close coordination with one another. However, there are a few key distinctions between a BIA and a risk assessment (**figure 2.29**):

- A BIA is used to establish requirements for business continuity efforts; this includes identifying resource dependencies, tracing those requirements to impacts and justifying investments that are associated with the disruption of services.
- A risk assessment focuses on understanding the threats faced by an enterprise, the frequency (likelihood/probability) and impact (loss magnitude/consequences) associated with those threats to properly respond to the risk, with the goal to decrease the likelihood or impact the organization could experience.

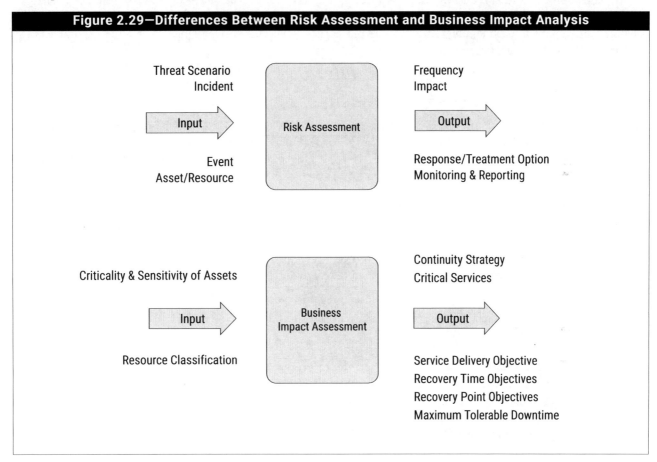

Figure 2.29—Differences Between Risk Assessment and Business Impact Analysis

2.9 Inherent, Residual and Current Risk

Risk is unavoidable in business. Risk is inherent in everything an enterprise does; however, some business processes may have a higher level of risk than others, and the degree of risk varies from one activity, product or service to another. The risk practitioner should understand risk and be able to assess and respond to any risk that lies outside the organizational risk appetite in a way that reduces it to an acceptable level. Accomplishing this task requires an ability to distinguish between inherent, residual and current risk.

Mature enterprises can leverage the concepts of inherent, residual and current risk in management's decision-making processes. However, when discussing risk, it is important to understand that it is a risk to the business; subsequently

CHAPTER 2— IT RISK ASSESSMENT

a risk practitioner should consider the current needs and readiness of the enterprise when communicating these concepts.

2.9.1 Inherent Risk

Inherent risk is the risk level or exposure without considering the actions that management has taken or might take (e.g., implementing controls). This is the risk that is ever present in any chosen course of action that is not specifically avoided.

For example, there is inherent risk in owning and operating an automobile. A person may be involved in an accident through no fault of their own. Having a vehicle readily available provides a number of benefits: easy transportation to work, the ability to take trips and vacations and freedom to run errands when needed. However, the risk of being involved in an accident will remain for as long as a person continues to own and operate the vehicle. This is considering that there are certain controls already in place, such as successfully passing driver's training and testing, antilock brakes and airbags, for example.

2.9.2 Residual Risk

Residual risk is remaining risk after management has implemented a risk response, which is typically a mitigation activity but may also include risk transfer. Residual risk is calculated by subtracting the effectiveness of the risk response (typically a control) from the inherent risk. If the residual risk associated with a particular control is within the organizational risk appetite, the residual risk should be accepted. Otherwise, an additional response is needed.

As another example, automobile insurance coverage can be increased, and additional safety features such as five-point restraints, lane change alerting, brake assist, forward-collision warning systems, pedestrian detection and automatic emergency braking can be added to reduce the impact if the risk is realized. Limits can even be placed on when and where the vehicle is driven to reduce the probability of an accident occurring. This should lower the residual risk but does not remove or alleviate the inherent risk of an accident because the threat is outside of a person's control.

2.9.3 Current Risk

When discussing the potential benefits of taking a specific risk response action (accept, mitigate or transfer), the risk practitioner typically uses the term "residual risk" in the predictive sense, referring to the remaining risk that will exist after a given response has been properly implemented. The usage of "current risk" tends to create confusion in organizations that do not have an established and mature risk management function, as this is the term used to describe the current "point in time" risk associated with an asset, where both actions taken and those still pending are taken into consideration.

Concluding the automobile example, current risk is the risk at a specific point in time, given the threat environment in which the asset (in this case, the car) is presently exposed. For instance, if the car is parked in a driveway, it is still susceptible to an accident (someone could pull into the driveway, not properly judge the distance and hit the car). However, given the car's location in a residential neighborhood and parked in a private driveway, the probability and impact would be a lot lower than if the vehicle is parked in a public parking lot or street.

---

[1] International Organization for Standardization/International Electrotechnical Commission (ISO/IEC); *ISO/IEC 27005:2018 Information Technology—Security Techniques—Information Security Risk Management*, Switzerland, 2018.
[2] Verizon, *2020 Data Breach Investigations Report*, USA, 2020
[3] The Open Web Application Security Project (OWASP), www.owasp.org
[4] ISACA, *Generating Value from Big Data Analytics*, USA, 2014
[5] ISACA, *Rethinking Data Governance and Data Management*, USA 2020, www.isaca.org/bookstore/bookstore-wht_papers-digital/whprdg
[6] SANDIA National Laboratories, *Categorizing Threat: Building and Using a Generic Threat Matrix*, USA, 2007

Chapter 3:
Risk Response and Reporting

Overview

Domain 3 Exam Content Outline 134
Learning Objectives/Task Statements 134
Suggested Resources for Further Study 135

Part A: Risk Response

3.1 Risk and Control Ownership 138
3.2 Risk Treatment/Risk Response Options 140
3.3 Third-party Risk Management 144
3.4 Issue, Finding and Exception Management 146
3.5 Management of Emerging Risk 148

Part B: Control Design and Implementation

3.6 Control Types, Standards and Frameworks 151
3.7 Control Design, Selection and Analysis 155
3.8 Control Implementation 156
3.9 Control Testing and Effectiveness Evaluation 161

Part C: Risk Monitoring and Reporting

3.10 Risk Treatment Plans 167
3.11 Data Collection, Aggregation, Analysis and Validation 170
3.12 Risk and Control Monitoring Techniques 175
3.13 Risk and Control Reporting Techniques 179
3.14 Key Performance Indicators 182
3.15 Key Risk Indicators 183
3.16 Key Control Indicators 186

CHAPTER 3— RISK RESPONSE AND REPORTING

Overview

By identifying and analyzing risk, enterprises become aware of what forces within and beyond their control might threaten their operations and how the effects of internal and external factors might manifest as impacts to production. However, documenting the results of this analysis in a risk register only provides awareness of the situation as it exists. From there, the next steps are for the enterprise to decide how to respond to the analyzed risk, which may include the design and implementation of new controls or enhancements to existing controls; and then to devise and conduct effective monitoring and reporting of risk to keep management informed in a timely manner.

This chapter represents 32 percent (approximately 48 questions) of the exam.

Domain 3 Exam Content Outline

A. Risk Response

1. Risk Treatment/Risk Response Options
2. Risk and Control Ownership
3. Third-party Risk Management
4. Issue, Finding and Exception Management
5. Management of Emerging Risk

B. Control Design and Implementation

1. Control Types, Standards and Frameworks
2. Control Design, Selection and Analysis
3. Control Implementation
4. Control Testing and Effectiveness Evaluation

C. Risk Monitoring and Reporting

1. Risk Treatment Plans
2. Data Collection, Aggregation, Analysis and Validation
3. Risk and Control Monitoring Techniques
4. Risk and Control Reporting Techniques
5. Key Performance Indicators
6. Key Risk Indicators
7. Key Control Indicators

Learning Objectives/Task Statements

Upon the completion of this chapter, the risk practitioner will be able to:

1. Collect and review existing information regarding the organization's business and IT environments.
2. Identify potential or realized impacts of IT risk to the organization's business objectives and operations.
3. Identify threats and vulnerabilities to the organization's people, processes and technology.
4. Evaluate threats, vulnerabilities and risk to identify IT risk scenarios.
5. Establish accountability by assigning and validating appropriate levels of risk and control ownership.
6. Establish and maintain the IT risk register and incorporate it into the enterprise-wide risk profile.

CHAPTER 3— RISK RESPONSE AND REPORTING

7. Facilitate the identification of risk appetite and risk tolerance by key stakeholders.
8. Promote a risk-aware culture by contributing to the development and implementation of security awareness training.
9. Conduct a risk assessment by analyzing IT risk scenarios and determining their likelihood and impact.
10. Identify the current state of existing controls and evaluate their effectiveness for IT risk mitigation.
11. Review the results of risk analysis and control analysis to assess any gaps between current and desired states of the IT risk environment.
12. Facilitate the selection of recommended risk responses by key stakeholders.
13. Collaborate with risk owners on the development of risk treatment plans.
14. Collaborate with control owners on the selection, design, implementation and maintenance of controls.
15. Validate that risk responses have been executed according to risk treatment plans.
16. Define and establish key risk indicators (KRIs).
17. Monitor and analyze key risk indicators (KRIs).
18. Collaborate with control owners on the identification of key performance indicators (KPIs) and key control indicators (KCIs).
19. Monitor and analyze key performance indicators (KPIs) and key control indicators (KCIs).
20. Review the results of control assessments to determine the effectiveness and maturity of the control environment.
21. Conduct aggregation, analysis and validation of risk and control data.
22. Report relevant risk and control information to applicable stakeholders to facilitate risk-based decision-making.
23. Evaluate emerging technologies and changes to the environment for threats, vulnerabilities and opportunities.

Suggested Resources for Further Study

Institute of Operational Risk, *Operational Risk Sound Practice Guide: Key Risk Indicators*, USA, 2010

ISACA, *Risk IT Framework, 2nd edition*, USA, 2020

ISACA, *Risk IT Practitioner Guide, 2nd edition*, USA, 2020

Muckin, Michael, S. C. Fitch, *A Threat-driven Approach to Cyber Security,* Lockheed Martin, USA, 2019

National Institutes of Standards and Technology, *Cybersecurity Framework version 1.1,* USA, 2018, https://www.nist.gov/cyberframework/framework

RiskLogix, "The Shortcuts Trap—Key Indicators Under Fire," 15 July 2019, https://risklogix-solutions.com/blog/key-indicators-under-fire

SELF-ASSESSMENT QUESTIONS

CRISC self-assessment questions support the content in this manual and provide an understanding of the type and structure of questions that have typically appeared on the exam. Questions are written in a multiple-choice format and designed for one best answer. Each question has a stem (question) and four options (answer choices). The stem may be written in the form of a question or an incomplete statement. In some instances, a scenario or a description problem may also be included. These questions normally include a description of a situation and require the

CHAPTER 3— RISK RESPONSE AND REPORTING

candidate to answer two or more questions based on the information provided. Many times, a question will require the candidate to choose the **MOST** likely or **BEST** answer among the options provided.

In each case, the candidate must read the question carefully, eliminate known incorrect answers and then make the best choice possible. Knowing the format in which questions are asked, and how to study and gain knowledge of what will be tested, will help the candidate correctly answer the questions.

1. Which of the following choices **BEST** assists a risk practitioner in measuring the existing level of development of risk-management processes against their desired state?

    A. A capability maturity model
    B. Risk management audit reports
    C. A balanced scorecard
    D. Enterprise security architecture

2. When a risk cannot be sufficiently mitigated through manual or automatic controls, which of the following options will **BEST** protect the enterprise from the potential financial impact of the risk?

    A. Insuring against the risk
    B. Updating the IT risk register
    C. Improving staff training in the risk area
    D. Outsourcing the related business process to a third party

3. When responding to an identified risk event, the **MOST** important stakeholders involved in reviewing risk response options to an IT risk are the:

    A. information security managers
    B. internal auditors
    C. incident response team members
    D. business managers

4. Which of the following choices should be considered **FIRST** when designing information system controls?

    A. The organizational strategic plan
    B. The existing IT environment
    C. The present IT budget
    D. The IT strategic plan

5. Which of the following choices is the **BEST** measure of the operational effectiveness of risk-management process capabilities?

    A. Key performance indicators
    B. Key risk indicators
    C. Base practices
    D. Metric thresholds

Answers on page 137

Chapter 3 Answer Key

Self-assessment Questions

1. **A. A capability maturity model grades processes on a scale of 0 to 5, based on their maturity. It is commonly used by enterprises to measure their existing state and then to determine the desired one.**
   B. Risk management audit reports offer a limited view of the current state of risk management.
   C. A balanced scorecard enables management to measure the implementation of strategy and assists in its translation into action.
   D. Enterprise security architecture explains the security architecture of an entity in terms of business strategy, objectives, relationships, risk, constraints and enablers and provides a business-driven and business- focused view of security architecture.

2. **A. An insurance policy can compensate the enterprise monetarily for the impact of the risk by transferring the risk to the insurance company.**
   B. Updating the risk register (with lower values for impact and probability) does not actually change the risk, only management's perception of it.
   C. Staff capacity to detect or mitigate the risk may potentially reduce the financial impact, but insurance allows for the risk to be completely mitigated.
   D. Outsourcing the process containing the risk does not necessarily remove or change the risk.

3. A. Business managers are accountable for managing the associated risk and determine what actions to take based on the information provided by others, which may include collaboration with and support from IT security managers.
   B. Risk response is not a function of internal audit.
   C. The incident response team must ensure open communication to management and stakeholders to ensure that business managers/leaders understand the associated risk and are provided enough information to make informed risk-based decisions.
   D. Business managers are accountable for managing the associated risk and will determine what actions to take based on the information provided by others.

4. **A. Review of the enterprise's strategic plan is the first step in designing effective IS controls that fit the enterprise's long-term plans.**
   B. Review of the existing IT environment, although useful and necessary, is not the first task that needs to be undertaken.
   C. The present IT budget is one of the components of the strategic plan.
   D. The IT strategic plan exists to support the enterprise's strategic plan.

5. **A. Key performance indicators (KPIs) are assessment indicators that support judgment regarding the performance of a specific process.**
   B. Key risk indicators (KRIs) only provide insights into potential risk that may exist or be realized within a concept or capability that they monitor, not necessarily at the process level.
   C. Base practices are activities that, when consistently performed, contribute to achieving a specific process purpose. However, base practices need to be complemented by work products to provide reliable evidence about the performance of a specific process.
   D. Metric thresholds are decision or action points that are enacted when a KPI or KRI reports a specific value or set of values.

CHAPTER 3— RISK RESPONSE AND REPORTING

Part A: Risk Response

The risk response phase of risk management, shown in **figure 3.1**, focuses on the decisions made regarding the correct way to address identified risk. Risk response is based on the information provided in the earlier steps of risk identification and risk assessment. The desire to reduce risk is balanced with the constraints placed on the enterprise through budget, time, resources, strategic plans, regulations, customer expectations and other business and industry factors. Management should be prepared to justify to stakeholders why its risk response decisions reflect the best balance of these competing factors. Additionally, an effective risk response process provides a road map to implementing agreed-upon changes and/or controls according to a reasonable schedule.

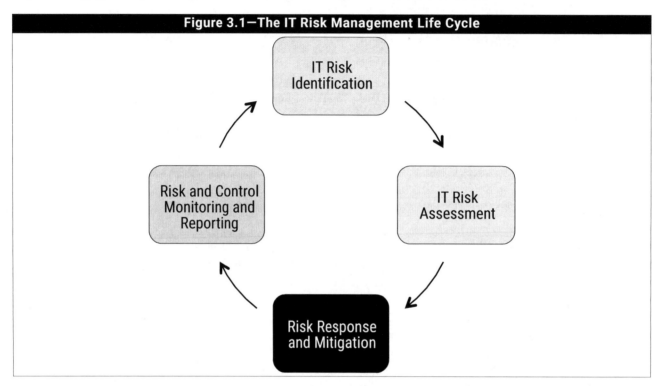

Risk response should be undertaken in a way that protects current operations to the greatest extent possible. In some cases, response activities may cause disruption. However, the goal should be to respond to risk without unduly impairing operations, and this expectation makes the careful selection of actions and controls to address risk an important part of the process.

3.1 Risk and Control Ownership

Every risk identified in the risk register should have a specified owner at a management level. This owner should be empowered to make decisions on behalf of the enterprise. In general, the risk owner should be a senior executive or manager responsible for one or more of the business processes or functions that may be impacted. The risk practitioner should communicate with risk owners to ensure that they are aware of what risk responses have already been implemented and responses that are pending implementation.

Current and anticipated levels of risk should be clearly conveyed in ways that provide insight into the aggregate risk that the organization as a whole is facing. Multiple low risk can aggregate to become moderate or even severe; for instance, the Global Financial Crisis of 2008 arose in part because numerous financial institutions had taken on an extremely large volume of investments with relatively low individual risk. Aggregate risk is evaluated against the strategic risk appetite and deals with organization-wide impacts. Where aggregation is incomplete or ineffective, senior executives may lack the information they need to make good business decisions to manage challenging circumstances. At its worst, the implications of poor aggregation can reach the level where an organization's continued existence may be in jeopardy.

3.1.1 Ownership and Accountability

The purpose of determining risk ownership is to ensure accountability is in place. A risk owner must be someone with the budget, authority and mandate to select an appropriate risk response based on analyses and guidance provided by the risk practitioner. If someone who lacks one or more of these is chosen as the risk owner, that person will be incapable of exercising the full scope of ownership responsibilities, and true accountability cannot be expected. Similarly, every risk should have a single risk owner. Where more than one senior leader may be impacted by a risk, seek consensus among those affected for which is the best suited to take ownership given resources, focus, expertise and other factors. Similarly, a control should be owned by the owner of the risk that it mitigates. However, the control may be owned by someone else in the case of controls that affect more than one risk. Risk and control owners should be recorded in the risk register.

In general, risk owners need to actively drive the risk management activities performed by risk practitioners to ensure the desired outcomes can be clearly understood by risk practitioners.

Accountability extends beyond the basic risk response decision to approving specific controls when mitigation is the chosen risk response. Types of controls are examined in more detail in section 3.6 Control Types, Standards and Frameworks. However, the concept is to create a direct link between risk and control, so that all risk is addressed through appropriate controls and all controls are justified by the risk that mandates their existence. This relationship is shown in **figure 3.2**.

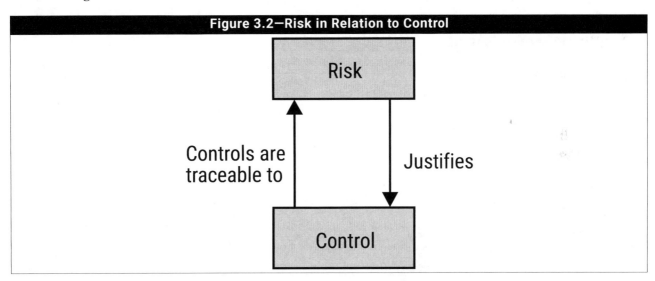

A detailed knowledge of controls may not be reasonable for all risk owners, and significant technical expertise is typically not necessary at the owner level. The risk owner generally owns any controls associated with that risk and is accountable for ensuring that their effectiveness is monitored. In some areas where there are regulations or laws that apply to risk, the risk owner may have to prepare (or direct preparation of) standard reports on the status of risk, any incidents that may have occurred and the level of risk currently faced by the enterprise from one or more functional areas, which can then be included in aggregate risk reporting across the enterprise.

3.2 Risk Treatment/Risk Response Options

The goal of risk response is to bring risk into alignment with the organizational risk appetite and tolerance. Factors such as on-staff expertise, strategic direction, legal and regulatory requirements and organizational culture should be considered when considering precisely what responses are best suited to particular areas of risk.

3.2.1 Aligning Risk Response with Business Objectives

The risk assessment report and risk register document the risk identified during the identification and assessment phases of the risk-management process. Both the report and the register should indicate the assessed level or priority of each risk. Risk practitioners should keep in mind that these recommendations are guidelines, and that several recommendations may be provided. Management is responsible for evaluating and responding to the recommendations included in the report, first by determining the best response and then developing an action plan and implementation strategy that addresses the risk in a manner consistent with the risk appetite and risk tolerance of the enterprise.

The enterprise exists to meet its mission in the manner deemed best by management. Accordingly, the decisions made around all aspects of risk management may diverge from recommendations made by risk practitioners. The role of the risk practitioner is not to decide on behalf of the organization. Instead, the risk practitioner's role is to ensure that management has the best available information to make informed decisions. Management should be kept aware of the drivers for risk management, such as compliance with regulations and the need to support and align the risk response with business priorities and objectives. This does not preclude a risk response that would negatively impact the ability of the organization to meet its mission, but it does mean that such a response would be considered very carefully.

3.2.2 Risk Response Options

Evaluation of an appropriate risk response is part of the risk-management process cycle, not a one-time activity. There are four commonly accepted options for risk response:

- Risk acceptance
- Risk mitigation
- Risk sharing/transfer
- Risk avoidance

Keep in mind that the purpose of defining a risk response is to bring risk in line with the enterprise's defined risk appetite and tolerance as cost-effectively as possible, not to eliminate or minimize the risk at all costs. All response activities incur some cost, typically a mix of the direct cost of response and the potential cost of impact. Finding the right balance is a management function that considers all of the constraints mentioned earlier, such as budget, time and resources.

Risk Acceptance

The choice to accept risk is a conscious decision made by senior management. Risk acceptance means recognizing both the existence of risk and its potential impact, then deciding with full awareness to allow (assume) the risk to remain without (further) action. Management is always accountable for the impact of a risk event should it occur, so the decision to accept a risk is made according to the organizational risk appetite and risk tolerance.

In some industries, risk acceptance may be more easily recognized by the term "self-insurance." Under a self-insurance model, an enterprise sets aside or has access to sufficient resources to absorb the cost of impact should a risk event occur and takes no further action. Self-insurance and risk acceptance are synonymous in practice.

Because the goal of risk management is to bring risk within acceptable levels as cost-effectively as possible, risk that falls within the organizational risk appetite without any action being taken generally should be accepted. Such risk is acceptable by definition, and acceptance of acceptable risk tends to be the most cost-effective choice. Exceptions to this approach may arise under unusual circumstances, such as cases where an action that would reduce an already acceptable risk would also result in increased business opportunity; however, such cases are rare.

Management may also choose to accept a risk *outside* of the risk appetite but within the range of the organization's risk tolerance. Risk tolerance is typically a factor when either there is no technically feasible remedy to the risk available or the cost of available actions outweigh their benefits. See section 1.10 Risk Appetite, Tolerance and Capacity for more information.

Examples of risk acceptance as a result of risk tolerance include:

- Projections indicate that a certain project will not deliver the required business functionality by the planned delivery date, but management decides to proceed with the project hoping that the delivery will come soon after the planned date.
- A particular risk would have catastrophic consequences but is assessed to be extremely rare, and approaches to reduce it are cost prohibitive. Management may decide to accept this risk.

It is imperative that risk be formally accepted by someone empowered to act on behalf of the enterprise at a level commensurate with the projected consequences associated with the risk. For a risk to be accepted informally or at the wrong level is not proper acceptance. Instead, this is a form of risk ignorance, which refers to both the failure to identify or acknowledge risk and the blind acceptance of risk without knowing or acknowledging what the risk level really is. Too often, acceptance of risk is based on poorly calculated risk levels. Many enterprises have found, in the wake of an incident, that the actual level of risk accepted was far greater than what they had intended to accept. Careful review of actuarial data or the outcomes of similar incidents at other enterprises and their resulting impact may help the risk practitioner to estimate true likely incident costs. Framing risk acceptance in terms of self-insurance may help risk practitioners explain the true implications of accepting risk to management.

In addition to accepting risk at the outset (which relates to inherent risk), enterprises should also recognize that other approaches to risk management typically yield residual risk that must be acceptable to the organization. A mitigated risk, for instance, is typically still non-zero and must be evaluated to ensure that it falls within the organizational risk appetite.

Additionally, risk acceptance is not a one-time process. The level of risk and potential impact of an incident can change dramatically as assets increase or decrease in value, new threats emerge, attackers gain additional skills or motivation, vulnerabilities are discovered, the existing controls become weaker or the operations team becomes less diligent. Enterprises should conduct regular reviews of their accepted risk to ensure that these continue to be consistent with senior management's levels of risk appetite and risk tolerance. Wherever a risk owner is unwilling to accept the current risk, a new risk response is necessary. This process is continued through iterations until the risk reaches an acceptable level, at which point the risk owner should formally accept the risk. Acceptance of residual risk should also include accountable ownership of the controls that produce the appropriate mitigation to ensure that these are suitably maintained and remain effective after acceptance.

Risk Mitigation

Risk mitigation refers to actions that the enterprise takes to reduce a risk. Some examples of risk mitigation are:

- Strengthening overall risk management practices, such as implementing or maturing internal processes

CHAPTER 3— RISK RESPONSE AND REPORTING

- Installing a new access control system
- Improving organizational policies or operational procedures
- Developing effective plans for incident response, disaster recovery or business continuity
- Using compensating controls to address areas of persistent weakness in production systems

Mitigation is typically achieved through the implementation of controls, which affect the frequency and/or impact of the risk. See Part C: Risk Monitoring and Reporting for more information.

Realistic choices for risk mitigation may be determined in part by the scope of the risk or the speed at which an incident can expand. In some cases, multiple controls may be needed to reduce risk to an acceptable level. Risk practitioners should also take care to account for the cost of controls relative to their effects on risk. Controls whose implementation cost exceeds the risk that they are meant to mitigate are cost-ineffective and should not be deployed without further justification.

Keep in mind also that the goal of mitigation is to reduce risk to an acceptable level, not to eliminate risk entirely. Once mitigation is complete, acceptance of residual risk should be documented and subject to periodic reevaluation in the same manner as the acceptance of inherent risk.

Risk Transfer/Sharing

As noted, risk acceptance can be a form of self-insurance. This differs from risk transfer, which is the decision to reduce loss by having another enterprise incur the cost. The most common example of risk transfer is the purchasing of third-party insurance, which provides a guarantee of compensation or replacement should a loss occur. Partnerships and outsourcing agreements are also forms of risk transfer, in which two or more organizations work together under an arrangement in which both risk of loss and potential for profit are divided among the participants according to agreed-upon terms and conditions. In general, the remedy for failures under an outsourcing or partner agreement tends to be monetary indemnity, so the effect of these arrangements is comparable to third-party insurance even though the legal and practical workings are somewhat different.

The term "risk transfer" is misleading, because an enterprise that sustains a substantial loss may accrue costs beyond what can be reimbursed by insurance, particularly in the form of damage to its brand (reputational risk). For instance, a major financial institution that exposes large amounts of confidential customer data might be reimbursed for immediate expenses by an insurance policy. However, customers are unlikely to be impressed by the insurance payout and may take their business elsewhere. These long-term costs are typically not covered by any form of risk transfer, because customers view their relationship as being with the enterprise and not its insurers, partners or outsourced services. Risk transfer is a legitimate option for responding to risk, but risk practitioners and managers should take care to remember that stakeholders rarely regard the transfer of risk as complete absolution of blame.

Transferred risk, like all other risk, should be reviewed on a regular basis to ensure that it remains appropriate and adequate. For example, an enterprise should ensure that the current amount of insurance is adequate to cover losses and that the enterprise is compliant with the terms and conditions of the coverage. With regards to outsourcing arrangements, effective management includes regular reviews of security compliance reports or operational procedures to ensure that the outsourced services are being done according to contractual agreements. Risk transfer is only effective when they are properly maintained, and the moment after an incident occurs is the worst time to realize that a contract recently expired, a premium was unpaid or the third-party vendor has not maintained agreed-upon business practices.

Risk Avoidance

In each of the three prior response options, the goal has been to create an acceptable level of risk under which to continue operations. In certain limited circumstances, it may be impractical or even technically impossible to bring the risk of an activity into line with the risk appetite and risk tolerance. Under such circumstances, the best choice may be risk avoidance, which means exiting the activities or conditions that give rise to that risk.

Risk avoidance is the choice that remains when no other response is adequate, meaning the following are true:

- The current exposure level is deemed unacceptable by management.
- Mitigation or transfer that would bring the risk in line with acceptable levels is either impossible or would cost more than the benefits that the enterprise derives from the activities.

Some examples of risk avoidance include:

- Relocating a data center away from a region with significant natural hazards
- Evacuating staff from a work area in which there is substantial civil unrest or threat of war
- Declining a project whose business case shows a substantial likelihood of failure
- Rejecting a partnership agreement in which potential losses are allocated to the enterprise, but the partner stands to benefit from most potential profits
- Deciding not to use proprietary technology or software from a vendor because it has only recently begun operations or faces uncertain future prospects

Consideration should also be given to the "costs" of risk avoidance. This cost can be measured in the loss of a business opportunity or other opportunities lost because of the decision to avoid the risk.

The actual decision to avoid risk is made by senior management. The role of the risk practitioner is to make the case for why avoidance may be the correct choice given the circumstances. For this reason, all other options should be considered and rejected before recommending risk avoidance, allowing the risk practitioner to explain in detail what else has been considered and why transfer and mitigation are unacceptable options.

3.2.3 Choosing a Risk Response

In some cases, the decision to pursue a particular risk response may not be immediately apparent. Mitigation and transfer may both be feasible options, or a risk may be acceptable but also have the potential for mitigation in a way that would improve operational efficiency. In such cases, the best risk response is typically determined based on analysis. Management may consider several factors, including:

- Priority of the risk as indicated in the risk assessment report
- Complexity of recommended controls in cases involving mitigation
- Alternatives that are suggested through further analysis
- Cost of the various response options, including:
    - Acquisition cost
    - Training cost
    - Impact of productivity
    - Maintenance and licensing costs
- Management decision based on:
    - Financial objectives
    - Tolerance and variability of loss

CHAPTER 3— RISK RESPONSE AND REPORTING

- Relative importance
- Risk tolerance and risk appetite
- Requirements for compliance with regulations or legislation
- Alignment of the response option with organizational strategy
- Possibility of integrating the response with other initiatives
- Organizational culture (see section 1.3 Organizational Culture for more information)
- Time, resources and budget available for the response
- Environmental and market conditions and evolution
    - Historical progress and transformation of the associated risk

The selection of the appropriate response is often based primarily on the value obtained for the cost. Where more than one risk response option would yield acceptable risk, does the cost of implementing a specific risk response provide enough value to the enterprise for it to be a wise decision? This is inherently a management decision. However, risk practitioners are often involved in framing the situation and answering questions.

Specific responses to risk are typically recommended in business cases, which document the problems and one or more options as well as providing detailed analyses on which management may base decisions. Beyond being common and useful tools for decision support, business cases also serve as operational tools against which to evaluate and support investments throughout their life cycles. The risk practitioner may find it helpful to conduct a cost-benefit analysis for each alternative to estimate its likely value to the organization relative to the immediate expense. Another option is to calculate return on investment (ROI), which can make it easier to see how much time would pass before the enterprise would recoup the cost through savings or increased revenue.

Risk responses should be documented in the risk register with clear accountability and time frames for completion. Many risk responses cannot be implemented instantly, and a plan of action and milestones (POA&M) can be useful as a means of identifying a basis for tracking progress and setting times of review up the point of completion. Risk practitioners should ensure that these reviews are completed and that planned mitigations are not superseded or abandoned to competing priorities.

3.3 Third-party Risk Management

Not all risk arises from within an enterprise. Outsourcing is a form of risk transfer, but reliance on a third-party provider also creates exposure. The risk practitioner should take care to address risk that arises when an organization outsources business functions, supportive services, IT services and data management. Typically, ownership of data and business processes remains with the outsourcing enterprise, not the contracted service provider. This relationship can create legal liability that may be difficult for the outsourcing enterprise to manage, because outsourcing by its very nature places most of the day-to-day operations, staff and procedures within the scope of the contract outside of the enterprise's direct control. The risk can be mitigated to a certain extent by the inclusion of carefully worded indemnity clauses that require the vendor to repay the losses suffered due to legal/regulatory violations on the service-provider side.

The relationship between an outsourcing provider and the enterprise is defined by a contract. Legal requirements written into contracts should address the jurisdiction for any complaints or disagreements with the provider, and regulations in the host country regarding disclosure to law enforcement and data transmission and storage across borders. The risk practitioner should advise that appropriate security and regulatory requirements are addressed in all agreements with suppliers and service providers. Service level agreements (SLAs) are also commonly included in outsourcing contracts, but the risk practitioner should make certain that risk owners do not misinterpret them. SLAs provide monetary remedies. Outsourcing does not terminate the ownership or liability of a risk owner for data or performance; for instance, the enterprise remains accountable for the security of information that it stores with a

CHAPTER 3— RISK RESPONSE AND REPORTING

third-party host. Standards of due care and/or due diligence may be warranted in certain situations or regulated industries.

When the management of data is outsourced, the outsourcing enterprise is responsible for ensuring that adequate security requirements and regulations for handling the information have been written into the outsourcing agreement. It is also critical that mechanisms exist to ensure that these requirements and regulations are being followed. Depending on the particular circumstances, an outsourcing organization may require the right to audit the processes of the outsource supplier or an attestation provided by external auditors, or an independent reviewer. Other issues related to outsourcing include declaring the jurisdiction of the agreement and which courts would hear any dispute related to the terms and conditions of the contract.

These arrangements should be included in the outsourcing agreement and given high priority, because they safeguard against potentially severe impacts.

Concerns that should be considered in relation to the risk of using an outsourced supplier include:

- Hiring and training practices of the supplier
- Reporting and liaison between the outsourcing enterprise and supplier
- Time to notify and respond to any incidents
- Liability for noncompliance with terms of the contract
- Nondisclosure of data or business practices
- Responding to requests from law enforcement
- Length of contract and terms for dissolution/termination of contract
- Location of data storage including backup data
- Separation between data and management of data of competing firms
- Existence and regular testing of resiliency plans (e.g., business continuity and disaster recovery)
- Handling of data after termination of contract (i.e., return, destroy, transfer, etc.)

When an enterprise is contracted to provide or deliver services or equipment, the risk of noncompliance with the agreement must be met through review, monitoring and enforcement of the contract terms. Any failure to meet contract terms must be identified and addressed as quickly as possible. Typically, ownership of data and business processes remains with the outsourcing organization, not the contracted service provider. This relationship can create legal liability that may be difficult for the outsourcing organization to manage, because outsourcing by its very nature places most of the day-to-day operations, staff and procedures within the scope of the contract outside the enterprise's direct control. Enterprises with influence over terms may want to ensure that third-party providers are obligated to provide them with notice of breaches that do occur, and summary reports of remediation efforts, at least for incidents that meet specified severity criteria.

Limited mitigation may be attained through the use of carefully worded indemnity clauses that require the vendor to repay the losses suffered due to legal or regulatory violations on the part of the service provider. Contractual terms are commonly enforceable through service-level agreements (SLAs). An SLA is a formal guarantee that certain performance targets or standards will be met and includes pre-determined compensation for failure to meet those targets. In many cases, compensation takes the form of monetary offsets against future expense. Risk practitioners should recognize that even generously crafted SLA remedies are a form of risk sharing and do not offset reputational or long-term impacts.

Where enterprises rely on third parties to carry out key aspects of their business functions, the need for third-party assurance takes on a different connotation. If a third party is going to perform work for the enterprise, nondisclosure agreements (NDAs) form the foundation needed to protect the intellectual property of the enterprise from being disclosed to unauthorized personnel. This is especially relevant when a penetration test or other assessment has been performed to ensure any uncovered vulnerabilities could not be exploited to attack the enterprise. Outsourcing services are also subject to disasters, even when they boast of highly redundant data center environments replicated across continents.

The risk practitioner should take steps to ensure any provider of outsourced services used by the enterprise is compliant with the organization's defined requirements or at a minimum generally accepted good practices, whether through external audits such as Statement on Standards for Attestation Engagements Number 18 (SSAE 18) or US Health Information Technology for Economic and Clinical Health (HITECH) reports or having the right to conduct an audit on behalf of the organization. Where possible, reviewing an outsourcing provider's security policies and procedures may provide an additional level of assurance. This can be difficult to arrange unless it is written into the original agreement, so the risk practitioner should be aware and involved in all discussions regarding the establishment of third-party agreements to ensure that executives make decisions with a clear understanding of the associated risk.

The risk practitioner should understand the variety of cloud computing models and capabilities which the organization may desire to pursue. These can include common offerings such as software as a service (SaaS), infrastructure as a service (IaaS) and platform as a service (PaaS) and new forms of cloud application development such as containers and dockers. The risk practitioner needs to be aware of where data have the potential to be stored, accessed and processed. Different jurisdictions where data are stored or processed may introduce an unacceptable level of risk in the forms of litigation or incident response efforts.

Despite the risk associated with third-party activities, outsourcing and contracting of specialized services are proliferating rapidly. The promise of leveraging outside capabilities that are superior to what can be easily built internally in a short time, so that the organization can focus on its core business and strategy without distraction, is a compelling value proposition.

3.4 Issue, Finding and Exception Management

Variation from the norm or desired outcome represents uncertainty and is therefore a source of risk. Enterprises can manage this risk by applying strong procedures to areas of variation, including formal approaches to configuration, release, exception, and change management. Despite these controls, issues and findings may arise, warranting special handling.

3.4.1 Configuration Management

Technical complexity and leading-edge technology use can be a key cause of information risk and scales with the size of an enterprise. Standardizing system configurations can greatly reduce this complexity by simplifying planning, testing, implementation and maintenance activities. The risk practitioner should first determine whether standard configurations have been established and approved and then verify their proper documentation. Effort should also be made to spot-validate that documentation is used when planning changes or rollouts of new software, and updated to reflect these changes upon their implementation.

3.4.2 Release Management

In environments where software is developed for internal use, formal management of releases of new updates or versions of developed applications is very important. Releases should be coordinated with production staff, subject to independent (non-developer) verification and protection prior to deployment and aligned with the enterprise's overall life cycle for system development (see chapter 4 Information Technology and Security for more information).

3.4.3 Exception Management

Policies and procedures exist to standardize decisions and behavior. However, enterprises may encounter circumstances where usual practices are unsuitable due to timely technical or administrative constraints. Any exception should be formally documented and approved by a senior manager or executive before it is acted upon. Every exception represents deviation from standards, and introduces complexity, which can create or obscure risk. Therefore, exceptions should be reviewed at least annually to ensure that they are still necessary, and any unnecessary exceptions should be removed.

3.4.4 Change Management

Requests to change systems or configurations should be subject to formal review and approval by a dedicated committee. Change tends to alter the risk profile of an enterprise, so effective risk management calls for change advisory boards (CABs) comprised of representatives from several business departments and responsible for overseeing all information systems operations and approving changes to those systems. Having a CAB with a sufficiently broad perspective provides a communications channel between the business units and the IT department, which helps to ensure that changes are proposed and considered in ways that cater to both technical requirements and operational efficiency.

Under an effective change control model, changes are submitted for review by the CAB, which verifies that:

- The change request does not unknowingly affect risk or security.
- The change is formally requested, clearly justified, approved and documented.
- The change is scheduled at a time convenient for the business and IT.
- All stakeholders affected by the change are advised of the change.
- The change request includes test, implementation and rollback plans.
- The change will not compromise the enterprise's security baselines.

The intention of this model is to provide a balance between allowing needed changes to occur and preserving system reliability and stability. CABs typically have procedures for "emergency" requests, but this should only take the form of changing the timing of when requests are considered. Allowing changes to be made to the production environment without proper CAB review may expose the enterprise to unidentified risk, including risk that may exceed acceptable levels.

The change management process interacts with and oversees all of the other processes previously described in this section. Updates to configurations, software releases and exceptions should all be processed as changes and subject to review by the CAB.

3.4.5 Issue and Finding Management

Internally identified issues and findings that arise from third-party assessments or audits demonstrate a need to take corrective action. However, before action can be taken, the issue or finding must be fully understood and prioritized relative to resources and other constraints. From there, the enterprise needs to determine what situation or outcome would constitute an acceptable end state and develop a response to create that end state. In most cases, responses will involve changes, which should go through the change management process previously discussed. Addressed issues and findings should then be monitored to ensure that the timely actions produced the anticipated results and identify any new issues that may arise as a result of the changes made.

3.5 Management of Emerging Risk

Risk is fundamentally dynamic because it arises from uncertainty. Even the most well-developed risk management program must operate on a continuous basis to be effective, because new areas of risk are always emerging. The deployment of new controls and introduction of new technologies both have the potential to contribute to emerging risk.

3.5.1 Vulnerabilities Associated with New Controls

Any changes that deal with controls should be carefully considered. This includes both the addition of new controls and modification of existing controls. Beyond their intended benefit to the enterprise, new or updated controls may also introduce new vulnerabilities. For example, an access control system that protects against unauthorized access may inadvertently deny access to legitimate users who forget passwords or lose access cards, resulting in calls for technical assistance, delays in processing and user frustration. Modified controls may also not perform as expected, potentially reducing their effectiveness. The risk associated with deploying a particular control may even exceed the risk that it is meant to mitigate, as can sometimes be the case with early adoption of unproven technologies or changes to procedures that eliminate quality-control functions in the name of efficiency. Enterprises should coordinate with stakeholders on a proactive basis and perform rigorous user acceptance testing (UAT) under conditions as close to real-world use as possible prior to full implementation of any additions or changes to the control baseline.

3.5.2 Impact of Emerging Technologies on Design and Implementation of Controls

Many enterprises take a proactive approach to the adoption and use of new technologies in an attempt to gain or maintain competitive advantage. Managing risk in a highly dynamic environment requires the risk practitioner to be alert to the development of new technologies and proactive in assessing the risk of their incorporation into the enterprise. In some cases, eagerness to bring in new technology may get ahead of policies put in place to govern the process of technical adoption, setting the stage for unknowingly accepting risk outside the risk appetite. This is of particular concern when adoption occurs informally (sometimes called shadow IT) or on the authority of local managers who are not empowered to accept the commensurate risk.

The risk practitioner may find it beneficial to use automated network scanners and other tools to detect unauthorized implementations of new devices or technologies that have not yet been reviewed and approved for use. Staff also should be made aware of the processes required to submit a new technology for consideration, and any exceptions to full evaluation should be documented and approved by management. People typically want to stay within rules, and where procedures to submit a technology for consideration are clearly communicated and fully understood, the risk of shadow IT tends to be less than in more opaque environments. This may be especially important in a time of digital transformation, where processes are increasingly automated or occur through networked means. Where alternate channels exist for data movement, organizations may find that their risk profiles have expanded far beyond expectation.

Emerging technologies can have significant effects on the design and implementation of controls, sometimes beneficial and other times detrimental. Controls that have previously been effective may be rendered obsolete with advances in computing power or improvements in algorithmic design. For instance, codes that were unbreakable 30 years ago may now be cracked using automated, randomized guessing (brute force) running on a mobile phone. Analysis of traffic patterns can help determine the best placement of network appliances and sensors or identify ways for an adversary to bypass these devices when used with malicious intent. Machine learning greatly simplifies the programming of certain types of monitoring devices simply by analyzing patterns within common-use network traffic; however, the risk practitioner should keep in mind that if malicious traffic is already present in an

CHAPTER 3— RISK RESPONSE AND REPORTING

environment during these initial phases of deployment, automated systems may learn to recognize it as part of the baseline.

It is important for the risk practitioner to maintain awareness of what new threats and attack vectors are gaining credibility in academic circles or visibility within real-world business environments. It is no longer feasible to rely on organizational obscurity or lack of clear adversaries as protection against malicious activity. Those who seek to compromise networks and systems for whatever reason seek opportunities to train and prove their skills. For this purpose, any enterprise may be a suitable target. Organizational risk profiles are also subject to sudden and potentially severe changes, as occurred with the global pandemic of 2020 and the shift of large portions of the workforce away from managed workspaces to home-office and remote environments. Controls need to be designed and implemented in ways that are suited to addressing the threats, vulnerabilities, and impacts of this dynamic environment. As conditions change, previous assumptions should be subject to regular re-examination by the risk practitioner, with the goal of identifying technical and procedural deficiencies before these can be exploited by waiting adversaries.

Page intentionally left blank

Part B: Control Design and Implementation

Mitigation is by far the most common response to risk. The risk practitioner should be aware of both the current control environment and any anticipated changes. Controls are implemented to reduce or maintain risk at acceptable levels; however, controls may be poorly maintained, unsuitable for the risk that they are meant to control or incorrectly configured. Controls should be regularly reviewed to determine and ensure their effectiveness.

In addition to evaluating the effectiveness of each control for its intended purpose, the risk practitioner should verify the correct balance between technical, administrative (managerial) and physical controls. Implementation of a technical control, such as a firewall, requires training for the staff member who manages or operates it, correct procedures for its configuration, assignment of responsibilities for its monitoring and schedules for regular testing. If these coinciding administrative controls are not in place, the technical control may be less effective, giving stakeholders a false sense of security and resulting in unidentified vulnerabilities, an ineffective use of resources and greater risk than anticipated or intended.

3.6 Control Types, Standards and Frameworks

Controls are generally categorized as belonging to one of five categories, described in **figure 3.3**.

Category	Description
Preventive	Inhibit or impede attempts to violate security policy and practices. Encryption, user authentication and vault-construction doors are examples of preventive controls.
Deterrent	Provide guidance or warnings that may dissuade intentional or unintentional attempts at compromise. Warning banners on login screens, acceptable use policies, security cameras and rewards for the arrest of hackers are examples of deterrent controls.
Detective	Provide warning of violations or attempted violations of security policy and practices without inhibiting or impeding these actions. Audit trails, intrusion detection systems (IDSs) and checksums are examples of detective controls.
Corrective	Remediate errors, omissions, unauthorized uses and intrusions when detected. Data backups, error correction and automated failover are examples of corrective controls.
Compensating	Offsets a deficiency or weakness in the control structure of the enterprise, often because the baseline controls cannot meet a stated requirement due to legitimate technical or business constraints. Placing unsecured systems on isolated network segments with strong perimeter security and adding third-party challenge-response mechanisms to devices that do not support individual login accounts are examples of compensating controls that, while not addressing the direct vulnerabilities, make it harder to exploit them.

Figure 3.3—Control Categories

The interaction between the control types is shown in **figure 3.4**. The time of control effects relative to an incident is shown in **figure 3.5**.

CHAPTER 3— RISK RESPONSE AND REPORTING

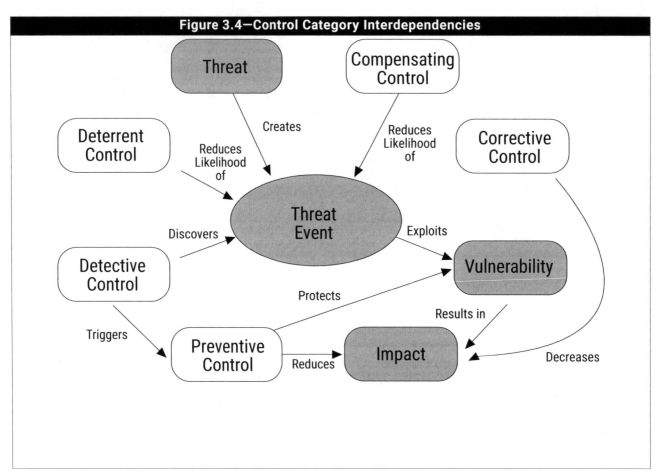

Figure 3.4—Control Category Interdependencies

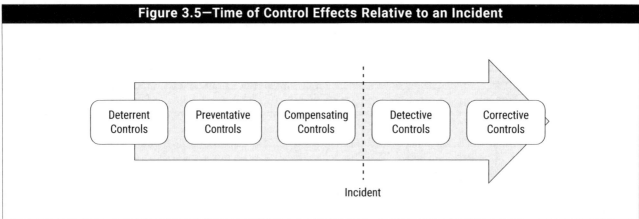

Figure 3.5—Time of Control Effects Relative to an Incident

Assessing the control environment provides the risk practitioner with an opportunity to evaluate both the risk culture of an enterprise and the effectiveness of its risk management program, which can be used to determine both the level of risk currently facing the organization and the seriousness of that risk. In general, risk is more serious when any of the following are true:

- Controls are inadequate or nonexistent.
- The wrong controls are being used.
- Controls are regularly ignored or bypassed.
- Controls are poorly maintained.

CHAPTER 3— RISK RESPONSE AND REPORTING

- Logs or control data are not reviewed or monitored.
- Controls are not tested on a regular basis or using standardized procedures.
- Changes to the configuration of controls are not managed.
- Controls can be physically accessed and altered by arbitrary individuals, without approval.
- Duties are inadequately segregated (see chapter 4 Information Technology and Security for more information)

3.6.1 Control Standards and Frameworks

Selection of controls requires evaluation and implementation. Based on data collected through appropriate analysis methods (e.g., cost-benefit analysis, ROI, etc.), management decides on the best available control or group of controls to mitigate a specific risk. A poorly implemented control may end up posing a risk to the enterprise if it creates a false sense of security or causes other controls to function less effectively. In particular, the implementation of technical controls requires proper procedures and training, and all controls should be assigned to specific owners accountable for ensuring that they are monitored and tested on a regular basis.

Adoption of standards and frameworks facilitates the process of selecting controls by specifying what should be done and, in the case of standards, also directing how to do it. Standards are written with the intent that adopting organizations will certify formal compliance. The 27000-series of standards from the International Standards Organization (ISO) focuses on information security, with ISO/IEC 27002:18 dealing specifically with security controls. Many industries also have standards for security within their respective sectors. For instance, the Payment Card Industry Data Security Standard (PCI DSS) is used by all organizations that process debit or credit cards. Similar standards are found in the healthcare, accounting, audit and telecommunications industries. Standards may be supplemented or even superseded by law or regulations in some locations.

In some cases, organizations may choose to adopt standards without intending to formally certify compliance. Standards may also include or be supported by frameworks, which provide sets of fundamental controls whose adoption is recommended, but whose implementation is left up to the organization. The U.S. National Institute of Standards and Technology (NIST) maintains a published Cybersecurity Framework (CSF) that adopts a threat-centric approach and breaks down recommended cybersecurity practices into Core components, rigor-influenced Implementation Tiers, and organization-specific Profiles. Even within the United States, CSF is voluntary guidance; however, organizations around the world use it as a source for leading practices when developing their cybersecurity programs. Similarly, the Cloud Security Alliance (CSA) has built a control framework tailored to the operational and security requirements of cloud computing.

Whether voluntary or mandatory, standards, frameworks and regulations all influence the particular control environments of organizations that adhere to them. See chapter 1 Governance for more information.

3.6.2 Administrative, Technical and Physical Controls

Controls can be further categorized as administrative (managerial), technical and physical, as shown in **figure 3.6**.

CHAPTER 3— RISK RESPONSE AND REPORTING

Figure 3.6—Control Methods	
Category	Description
Administrative (managerial)	Related to the oversight, reporting and operations of a process. Typically, performed by people rather than automated, these include controls such as policy and procedures, training and awareness, configuration/change management, employee development and compliance activities. Administrative controls tend to be subject to considerable human judgment.
Technical	Provided through the use of digital technology, equipment or devices. Sometimes called "logical" controls, technical controls include firewalls, network or host-based intrusion detection systems, passwords and antivirus software and generally safeguard networks, computer systems, and data. A technical control requires proper administrative controls to operate correctly and can generally be compromised by someone who gains physical access to its source of activity.
Physical	Based on restriction of physical access to something, which may be a particular device, room or facility. Physical controls include locks, fences, armored conduit, security guards and other factors that interact with the physical world and monitoring capabilities that are physically directed (such as closed-circuit television cameras). Physical controls require maintenance, monitoring and the ability to assess and react to an alert should a problem be identified.

3.6.3 Capability Maturity Models

The risk practitioner may find it useful to compare the state of the enterprise's risk management program to an established model of capability maturity that contains the essential elements of effective processes for one or more disciplines. A capability maturity model (CMM) describes an evolutionary improvement path from *ad hoc*, immature processes to disciplined, mature processes with improved quality and effectiveness. With the aid of a maturity model, enterprises can set realistic long-term goals for risk management by having a clear understanding of their current maturity (in terms of current working practices) and the areas that require improvement.

A mature enterprise that has defined, reliable processes that it follows consistently and continuously seeks to improve is much more likely to prevent incidents, detect incidents sooner and recover rapidly from incidents. As a general rule, enterprises should seek to implement well-structured risk management procedures across all departments and regions and ensure that every business process, system and software development project follows core risk management principles, policies, procedures and standards.

Key elements that may be used to measure the maturity of IT risk management capability include:

- Support of senior management
- Regular communication between stakeholders
- Existence of policies, procedures and standards
- Availability of a current business impact analysis
- Logging and monitoring of system activity
- Regular review of logs
- Scheduled risk assessments and review
- Testing of business continuity and disaster recovery plans
- Training of staff
- Involvement of risk principles and personnel in IT projects
- Gathering feedback from users and stakeholders
- Validating the risk appetite and risk acceptance levels
- Time to detect/resolve a security incident
- Risk mitigation strategies
- Standardized measures for assessing risk and defining risk ratios

- Exception management
- Risk and control ownership

Improvements in both the efficiencies and effectiveness of the risk management capability are based on the consistent application of policies and procedures. If these procedures are followed on a consistent basis, then the efficiency and effectiveness of the risk management practices will improve. The levels defined by a CMM are typically attained on the basis of self-assessment, although certain CMMs do provide a means for third-party assessment and validation.

3.7 Control Design, Selection and Analysis

When designing controls, the first step is to be aware of the current state of the control environment. The risk practitioner determines the current state of IT risk using the results of control testing activities and incident management programs. A current-state assessment therefore refers to the condition of the program at a point in time and should be regarded as accurate only at the time the risk state is measured. However, it does accurately reflect the current state, and regular reviews may be used to determine the state over a defined time frame.

Current risk levels are essential because they provide the reference point to understand the gap between the current and desired state. This gap should be investigated to determine its reason and to identify workable solutions that may address the disparity. Some tools that can be used by the risk practitioner to determine the current state of IT risk include:

- Audits
- CMMs
- Control tests conducted by the control owner or custodian
- Incident reports
- IT operations and management evaluation
- Enterprise architecture assessment
- Logs and automated monitoring
- Media reports of new and emerging vulnerabilities and threats
- Direct observation and self-assessments
- Third-party assurance
- User feedback
- Vendor reports
- Vulnerability assessments and penetration tests

Although an audit report tends to provide the most comprehensive view of the current state of controls, risk practitioners should take care in requesting audits. Auditors rely on evidence and use of published, rigorous methodology which leaves little room to argue with findings, which some managers and risk owners see as negative appraisals of their ability. Where use of auditors is permitted, their assistance can be invaluable. An alternative to requesting an audit is to obtain a copy of a recent audit report and use it as the starting point for self-assessments to identify any changes in the environment.

3.7.1 Control Design and Selection

Once the specific shortcomings of existing controls and configurations are understood, the next step is to design or select controls or enhancements to close the gap between the current state and an acceptable level of risk. Controls may be proactive—meaning that they attempt to prevent an incident or reactive—meaning that they facilitate the

CHAPTER 3— RISK RESPONSE AND REPORTING

detection, containment and recovery of operations should an incident occur. Because these are different roles, proactive controls are sometimes called safeguards, while reactive controls are often called countermeasures. For example, a sign that warns a person about a risk of fire is a safeguard against a fire starting, while a fire extinguisher or sprinkler system is a countermeasure that puts one out should it start.

Every enterprise has controls. In general, an effective control is one that prevents, detects or contains an incident, or enables recovery from a risk event. However, the purpose of a control is specifically to maintain risk at an acceptable level. That a control functions as intended does not unto itself guarantee that it is effectively managing risk, especially as the enterprise and its business processes change over time.

It is common for an enterprise to have some situations where the controls currently in place are not sufficient to adequately protect the enterprise. In most cases, this requires the adjustment of the current controls or the implementation of new controls within the system, referring to deterrent, preventative, detective or corrective controls. In some cases, it may not be feasible to reduce the risk within the system to an acceptable level by either adjusting or implementing controls due to reasons such as cost, job requirements or availability of controls. For instance, a small enterprise that has only a single individual with administrator rights generally cannot implement proper segregation of duties (SoD). Industrial systems, sometimes called operational technology (OT), are also often incapable of supporting additional controls due to custom integration and nonstandard protocols and architectures. In such cases, the risk owner may choose to implement compensating controls to reduce the risk. Compensating controls address weaknesses through concepts such as layered defense, increased supervision, or increased audits and logging of system activity. Compensating controls must be effective and protect the identified assets or processes. These must not be duplicated, redundant or increase the existing risk surface. These measures combine with existing controls to offset the risk that could not be addressed directly.

Controls are typically selected from capabilities either already present within the environment or available commercially. However, controls may be designed specifically for environments based on need. Where design is the intent, the risk practitioner should consider not only functionality but also maintenance and sustainability. Whether controls will be automated or manual, fail-safe (open) or secure (closed) or provide inputs to other controls or processes are all factors that should be considered. Designers should also understand throughput or speed requirements of associated or affected processes to ensure that any delay or impediment to regular processing remains within tolerable levels. These same considerations apply when analyzing existing controls within operational systems.

3.8 Control Implementation

Selected controls should be tested in an environment distinct from the production environment before implementation. To the greatest extent possible, this test environment should mimic the production environment, including as many instances of existing controls in their current states as possible. In prior years, establishing such test environments tended to be extremely costly. Today, with virtualization and container technology, it may be relatively trivial to establish a testing environment, even as a temporary arrangement to evaluate specific capabilities.

Successful testing of new or enhanced controls in as realistic a context as possible is a necessary precondition before committing to a changeover. The changeover process should be coordinated through formal change management, with any system outages communicated to all likely stakeholders across the organization. Where possible, deconflict change implementation with critical business functions to avoid impacts in the event that an outage goes longer than expected or the changeover is not successful. In general, this deconfliction occurs within the CAB as part of the change management process.

Implementation and testing should not be limited to technical controls. Administrative and physical controls require comparable levels of coordination, rigor and evaluation before being fully implemented and relied upon to deliver intended outcomes. Additionally, technical controls affecting access permissions to networks, systems or data are

CHAPTER 3— RISK RESPONSE AND REPORTING

frequently implemented according to segregation of duties (SoD) determinations made on an administrative basis. Wherever possible, changes in policy or procedure should be promulgated in draft format and assessed for operational impact across a diverse subset of the affected users before becoming mandatory.

3.8.1 Changeover (Go-live) Techniques

After final tests have been conducted and change management has been coordinated, implementation begins as scheduled. There are three common changeover methods used to transition from one system, technology, application or version to another:

- Parallel
- Phased
- Abrupt

Additionally, every changeover should account for the possibility of a fallback (rollback) scenario. These concepts are discussed below in detail and apply equally to basic controls within complex systems as to changes made at the system level.

Parallel Changeover

A parallel changeover means operating both the new system and the old system simultaneously. Parallel processing allows the project team greater latitude to test the reliability and performance of the new system and ensure that it is working correctly before removing the old system from service, which minimizes the risk of a failed changeover to the new systems and allows staff to train on and become familiar with the new system before it is in full production.

Parallel changeover provides the safest and quickest means of changeover in case the changes need to be rolled back. Since the current system is fully operational, the new system can be stopped at any point where results are not as intended.

Disadvantages of a parallel changeover include the cost of maintaining and monitoring both systems at the same time and ensuring that the data are consistent between both systems (because both are operational).

Figure 3.7 shows the principles of a parallel changeover.

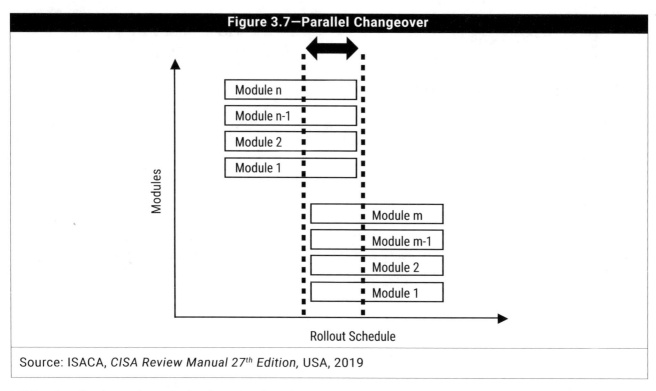

Figure 3.7—Parallel Changeover

Source: ISACA, *CISA Review Manual 27th Edition*, USA, 2019

While virtualization reduces the hardware burden of this arrangement, parallel processing is true duplication of effort and typically doubles licensing costs relative to normal operations. In some environments, it may be possible to operate in partial-parallel by limiting duplication to key functions. As a general rule, the cost of parallel processing is justified where there is significant complexity that may lead to an extended outage *and* the cost of that outage exceeds the cost of operating in parallel. For instance, when adopting a new enterprise-wide authentication system, maintaining an ability to authenticate under the old system for a period of time may be justified, because authentication affects all functions and failure could constitute a self-imposed denial of service against the entire enterprise. In contrast, if the control has a limited scope, a parallel approach may be too costly.

Phased Changeover

A phased changeover is conducted by replacing individual components or modules of the old system with new or modified components. This reduces the risk by gradually rolling out the new modules in ways intended to avoid impacting the entire system. Depending on the type of system, this type of changeover is not always possible. Some risk factors that may exist in a phased changeover include:

- Resource challenges arising from having to maintain unique environments with distinct hardware or software
- Operational challenges arising from having to maintain and monitor two different sets of user guides and procedures and potential confusion over changes in terminology
- Challenges maintaining consistency of data
- Extension of the project life cycle to cover delays that may arise from incremental changes not going as planned
- Change management for requirements and customization at each phase such as quantity, frequency and SLA or operational level agreement requirements

Figure 3.8 shows the principles of a phased changeover.

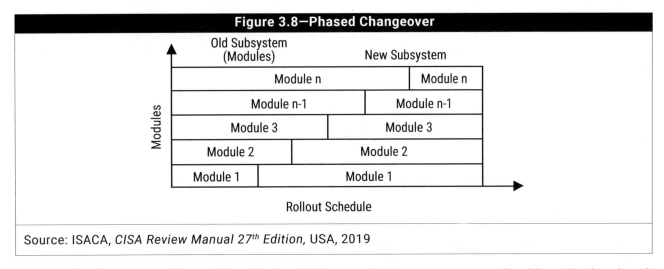

Figure 3.8—Phased Changeover

Source: ISACA, *CISA Review Manual 27th Edition*, USA, 2019

Some systems that appear to be candidates for phased changeover may present unexpected problems. Testing phased implementations is difficult, because the test environment may not accurately represent the state of production as it moves further from the starting point. Modules that are expected to deliver the same outputs may differ in ways that are sufficient to have downstream effects, resulting in impacts to data integrity or system performance. For instance, a logging system that stores timestamps in a different format may create errors in automatic parsing software or create false-positive responses in intrusion prevention systems that have already established heuristic baselines.

When implementing changes using a phased approach, it may be beneficial to build in considerable leeway at each phase to take full account of how the project is progressing. This is especially valuable in complex systems, where the effects of a change may not be known until the full production calendar has completed a cycle.

Abrupt Changeover

An abrupt changeover means single-instant movement from the old system to the new system, with the old system immediately taken offline. Abrupt changeover is the riskiest of the changeover processes because of the potential for lost opportunities in business processing if it becomes necessary to revert to the earlier system. For this reason, abrupt changeover is most prevalent in cases in which rollback is relatively assured or the impact of lost processing is minor. Many limited-scope controls are implemented using an abrupt changeover model.

Figure 3.9 demonstrates the principles of an abrupt changeover.

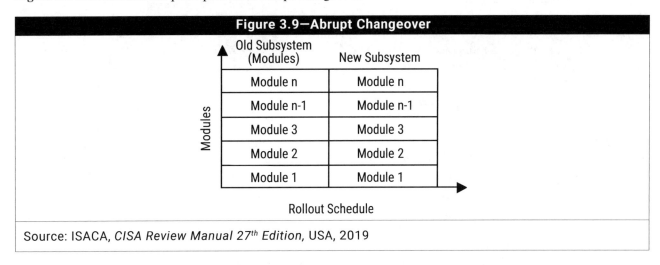

Figure 3.9—Abrupt Changeover

Source: ISACA, *CISA Review Manual 27th Edition*, USA, 2019

CHAPTER 3— RISK RESPONSE AND REPORTING

Although the technical element of an abrupt changeover happens suddenly, this should not be interpreted as a lack of preparation. Updates to documentation, staff training, and communication to stakeholders should all happen in advance of the cutover. In fact, it is more important that these activities be done deliberately and in advance in an abrupt changeover than in a parallel or phased scenario. Risk practitioners coordinating abrupt changeovers should ensure that the enterprise has adequate support capabilities available to handle potential disruptions and rollbacks in a timely manner.

Challenges Related to Data Migration

Migrating from one system or making fundamental modifications to an existing system may require the migration or conversion of data, which poses inherent risk to both data integrity and availability. The risk practitioner should assess the process used for data migration to identify any likely areas of concern. Additionally, procedures for recovery of data in its pre-conversion state should be developed, tested and practiced before conversion as insurance against unexpected negative outcomes. Some considerations for data conversion are listed in **figure 3.10**.

Figure 3.10—Data Conversion Key Considerations	
Consideration	Guidelines for the Risk Practitioner
Completeness of data conversion	Verify that the total number of records from the source database appears in the new database without loss (minus any intentionally removed).
Data integrity	Verify that the data are not altered manually, mechanically or electronically by a person or program, except strictly as required by conversion. Integrity problems include transposition and transcription errors, substitution or overwriting of records, and problems transferring particular records, fields, files, libraries and instances.
Storage and security of data under conversion	Verify that data are backed up before conversion and that controls are in place to safeguard data while in storage or transit. Unauthorized or unnecessary copies may lead to misuse, abuse or theft of data from the system.
Data consistency	Verify that the same field/record accessed by the new application is consistent with that used by the original application for the same purpose, which ensures consistency during repeat testing.
Business continuity	Verify that the application supports added or appended records, which is important for ensuring seamless business continuity.

Fallback (Rollback)

In any change scenario, whether it involves implementing a new system or making changes to an existing system, there is always a possibility that the change will not work as anticipated. Regardless of the cause of failure, the project team should have a fallback plan so that it is possible to return to the pre-change status and resume normal processing. The procedures for this reversion should be developed and approved as carefully as the change plan and subjected to rigorous testing before the changeover is attempted.

3.8.2 Post-implementation Review

Whether an implementation project succeeds or fails, a timely post-implementation review offers the best opportunity to capture lessons learned so that they can be applied to future projects. Some questions to ask during the review include the following:

- What went well during the project, and what could have been done better?
- Did the project bring the risk within acceptable risk levels?
- Were all user requirements and business objectives met?
- If any inadequacies or deficiencies have been identified, how might these be addressed now?

- Were specified methodologies, standards and techniques followed? If not, why not?
- How does the final cost compare to estimates?
- Were the project targets accomplished in terms and resources or were additional ones needed?

It is considered a good practice to hold a post-implementation review as soon as is practical after completing an implementation project. However, not all lessons associated with a given project may be immediately evident. A second, joint review by the implementation team and key stakeholders is often worthwhile after the project has been completed and changes have been in production for a sufficient time period to assess their effectiveness and value to the enterprise. In addition to highlighting unexpected deficiencies, this second-pass review may also reveal unanticipated value.

3.8.3 Control Documentation

The risk practitioner plays a key role in ensuring that controls are properly set up, operated, maintained and evaluated, with results reported to management on a regular basis. Formal documentation promotes consistency in task execution that manifests in reliable reports. Control management procedures include:

- Proper installation
- Creation of policies and procedures to support operations
- Implementation of change management procedures to ensure correct configuration
- Training of staff to monitor, manage and review controls
- Assignment of responsibility for monitoring and investigation
- Creation of a schedule for review and reporting
- Measurement of initial key performance indicators (KPIs)

As risk responses are implemented, the risk practitioner should ensure that each mitigation project has been implemented according to the intent and design of the project architects and that any changes did not erode or diminish effectiveness. Each control should have its own documentation that includes justification for its implementation, accountable and responsible owners, and review and reporting schedules. Exclusions applicable to the functioning of a particular control, such as omitting certain kinds of traffic or allowing certain roles to bypass it, should also be carefully justified and documented.

3.9 Control Testing and Effectiveness Evaluation

Controls exist to mitigate risk. When improperly implemented, the opposite may occur; not only may a control fail to meet business requirements, poor implementation or functional inadequacy may result in a security breach. Control testing is an opportunity to uncover flaws early enough to prevent that failure and do so cost-effectively. A comprehensive testing program includes testing at as many levels as may be needed to assess the complete scope of implementation. This includes testing at unit/component, integration/system and end-user levels, good practices based on standards used at each level and the results of all tests formally documented. Controls should be tested in ways that account for their full intended scope as well as any applicable exclusions.

Testing may be done on a progressive or regressive basis. Progressive testing begins with expectations and looks for flaws, while regressive testing works backward from known problems to identify causes. Both forms of testing are commonly used throughout the testing process. The risk practitioner may find regression testing to be useful in determining whether incidents have root causes in policy or standards.

3.9.1 Good Practices for Testing

Effective testing typically considers specific considerations for data, version control and code to the extent that these apply to particular controls. These are discussed in the following sections.

Data

The validity of testing depends, in large part, on the appropriateness of test data. All sensitive data elements should be displayed in a manner that makes them unreadable, such as masking sensitive information by displaying it as special characters or altering the data itself. Test data should be complete and allow the testing of all possible process functions and error handling, which includes "fuzzing"—testing the limit of the acceptable range of values and values beyond the allowable range in order to verify the functionality of input validation and process integrity controls.

Although testing simulates production use, the behavior of a system being tested is typically not fully predictable. Using distinct test data rather than production data is preferable in order to prevent the disclosure of sensitive data to people or systems that may not be approved to view or process it. Production data should be used as test data only in exceptional cases upon successful de-identification and specific management approval.

Environmental Separation

As mentioned previously, development or testing environments should be separated from production areas to prevent the potential for cross-population of data or application code outside the approval process. Depending on the enterprise, this separation may be done at the network level, may include physically separate workspaces, or may require entirely distinct teams of developers to be working on test and production systems. The risk practitioner should understand the separation model used by the organization and the rationale for managing separation in that fashion to ensure that it is consistently applied in practice. Wherever physical separation, logical separation or both are difficult, the enterprise should implement compensating controls commensurate with the risk level.

Version Control

There is always risk that a change to a system will overwrite or bypass functionality previously implemented or changed in a system. Version control refers to the assignment of specific version numbers or tracking mechanisms for each revision of a system, making it possible to distinguish between versions. Version control tracks the current iteration of something in development, checks for previous changes when a new change is implemented and can be used to ensure that the correct source version is staged for a move to production use. The risk practitioner should take care to ensure that reports from content reviews (including code testing) and other testing apply to the correct versions of components and systems.

Code Locking

Part of the process of moving a new or modified system or application into production is to lock down the source code in order to ensure that it cannot be inadvertently modified or intentionally tampered with after it has been approved for final testing. Some organizations call this final version gold code or locked code. In high-sensitivity environments, integrity-checking tools such as digital checksums may be used to validate that no changes have been made since approval.

CHAPTER 3— RISK RESPONSE AND REPORTING

Unit Testing and Code Review

Systems function on the basis of the combined effort of their components. Because complexity increases potential points of failure, it typically becomes more difficult to pinpoint the sources of problems as components are integrated. Testing done at the level of a particular control, module or section of code cannot demonstrate full operational capacity within a system. However, it may reveal certain errors in ways that can be remedied before moving to integration.

In software development environments, developers follow organizational coding standards, including proper use of comments, and these standards should reflect good practices of the industry wherever compatible with organizational goals. Additionally, all new or changed source code should be reviewed by a third party to validate quality and compliance prior to production use. The same basic standard applies to administrators making changes to configurations for routers and other network hardware. Third-party review is valuable because it can detect both unauthorized changes and implementations of error handing, input validation or documentation that may be inadequate. The process of third-party review should be done by an impartial, knowledgeable party and should be based on standards, not personal preferences.

Self-assessments are valuable as part of the development process. However, formal testing should always be done by individuals not involved in the creation of the deliverables being tested. Where detailed engineering information is available, it may be possible to perform this testing from a vulnerability assessment perspective. Commercial software and vendor products typically do not provide sufficient insight into their fundamental operations to allow for this informed perspective, making a review a more subjective assessment based on the penetration-testing model.

It is unlikely that the risk practitioner will be expected to conduct unit testing or code reviews, but he or she should be aware of the concepts and limitations of each approach and ensure that standards are in place and are used as the basis for any testing that is performed. In addition, the risk practitioner should be aware of any organizational or departmental policies that apply to testing and ensure that procedures are designed in ways that account for and support these higher-level controls.

Integration/System Testing

After the individual components of a system have been tested, the next step in a complex system is for the entire system to be tested to assess how the components work together with their interfaces and deliver overall operational capability. Integration testing is distinct from component-level testing because it shows the interactions of the system with upstream and downstream peers. System testing should be performed whenever a new control is implemented or an existing control is altered, because controls may have effects beyond their anticipated scope.

Unit testing and initial integration testing are often performed in a separate area from final system testing in order to separate the developer from the final testing and acceptance process and help ensure that the components being tested are not being modified by the developer during the testing process. This separation also provides a level of assurance that where system problems arise, they are integration problems and not problems with the individual components.

One purpose of system testing is to assess the security functions designed into the system and ensure that these have been built according to the design (considering any deviations that have been properly approved and documented). **Figure 3.11** describes different testing options available at the system level.

CHAPTER 3— RISK RESPONSE AND REPORTING

Figure 3.11—Options for System Testing

Test	Description
Recovery	Checks the system's ability to recover after a component failure
Security	Verifies that the modified/new system includes provisions for appropriate access controls and does not introduce any vulnerabilities that may compromise other systems
Stress	Determines the maximum number of concurrent users/services the application can process by increasing the number of users/services on an incremental basis
Volume	Determines the maximum volume of records (data) that the application can process by increasing the volume on an incremental basis
Stress/volume	A hybrid approach that uses large quantities of data to evaluate performance during peak hours
Performance	Compares the performance of the subject system to similar systems using well-defined benchmarks
Compliance	An approach for compliance-specific regulations or standards (such as the Global Data Protection Regulation) that would prevent personally identifiable information from remaining in the test environment

User Acceptance Testing

For a project to deliver what it promises is a significant accomplishment. However, a project whose deliverables are not what the enterprise needs cannot be called successful beyond a semantic level. Processes by which requirements for projects are developed are notorious for misalignment with functional capabilities that are needed in practice, and the best way to identify these areas of conflict is to allow the intended users of a system to interact with it as early as possible through UAT.

The purpose of UAT is to verify that the system meets user requirements and expectations, not whether it meets the stated design. UAT may highlight problems with functionality, training or process flow unanticipated by the requirements process and therefore undetectable by unit or integration testing. A failure in UAT suggests flaws in the enterprise's processes for needs analysis and requirements definition. For this reason, IT project management methodologies are increasingly incorporating UAT into the development process on an iterative basis, so that key functionality can be assessed early enough that changes may be considered and approved without negatively affecting the project schedule.

Quality Assurance

Quality assurance (QA) is a systematic plan of all actions necessary to provide adequate confidence that an item or product conforms to established technical requirements without the need for testing. When these actions are carried out in a tightly controlled production environment, the result can be a very high degree of reliability that what was produced necessarily conforms to expected standards. A failure in QA suggests flaws in the organization's processes for development and execution, and the basic QA paradigm includes detection of such flaws and their remediation in the assurance processes to eliminate repetition of problems in future iterations.

QA addresses whether the project delivered what it promised in the stated design, not whether that design actually meets user requirements and expectations. Accordingly, even the most mature QA process cannot replace UAT. Customization outside prescribed standards also tends to work against QA, which is most effective where the target of production is an exact replica of an ideal model.

Testing for Non-technical Controls

Administrative and physical controls should be tested with the same level of rigor applied to technical controls and may be assessed from the same progressive or regressive perspectives. Many physical security systems have known vulnerabilities that can be reduced or eliminated through effective simulation training, such as identifying blind spots in camera coverage or distorted images due to insufficient lighting.

Performance of security functions by humans should be assessed against clear procedures on the basis of adherence to expectations and norms; however, the foremost reason for employing human security is to afford an opportunity to exercise judgment that a technical control cannot demonstrate. Because every exercise of judgment represents a deviation from expected outcomes, human security performance should be assessed under conditions as realistic as is practical, with sufficient frequency to establish operational norms.

In workplaces where information is central to the business, the challenges associated with security roles provided by humans extend beyond uniformed guards. Analysts, operators and administrative staff are frequently in positions where action or inaction may create or increase risk. Policies and procedures are the means by which organizations inform human judgment to keep risk within acceptable levels. As with traditional security functions, adherence to these procedures and overarching policy should be tested under realistic conditions whenever practical. These assessments may include spot-checks of vital functions, structured walkthroughs of business processes and statistical evaluation of incidents based on human error.

3.9.2 Updating the Risk Register

The risk register should show the progress of testing and attainment of milestones during the progress of each mitigation project. After a control has been validated as effective, residual levels of risk are formally accepted by risk owners, with the risk register updated to reflect the changes. By keeping the risk register accurate and up to date, the risk practitioner ensures that it is consistently available as a resource for risk management activities across the enterprise.

Page intentionally left blank

Part C: Risk Monitoring and Reporting

Enterprises rely on monitoring and reporting functions to identify risk for assessment and response. The risk practitioner can best manage risk when monitoring is broad enough to provide a reasonable view of the risk environment but not so broad that the results are lost in a flood of data. Indicators for both performance and risk should be carefully considered and deliberately chosen based on their alignment with organizational goals. Identification and use of key risk indicators (KRIs), performance (KPIs) and controls (KCIs) can greatly improve the process of monitoring. In addition, periodic assessments and testing may also be necessary as a means of identifying new and emerging risk. Because of the changing nature of risk and associated controls, continuous monitoring is an essential step of the risk management life cycle as seen in **figure 3.12**.

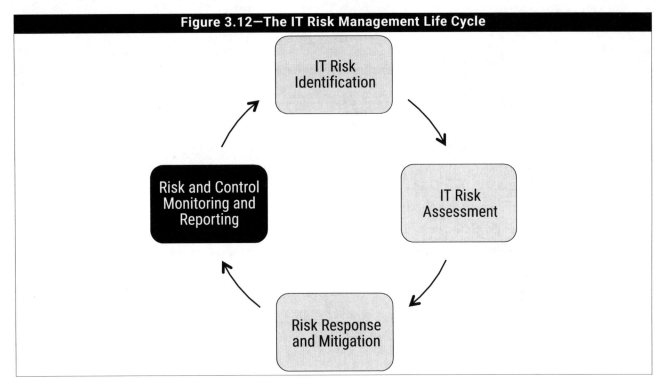

Although monitoring is essential, its effectiveness depends in large part on its successful integration with reporting. Consistent, repeatable methods of reporting provide management with a means of assessing the efficacy of the risk response and mitigation activities and justify the expenses of supporting security controls. Reports also assist management in exercising due care and due diligence in protecting the assets of the organization and meeting regulatory requirements.

As the goals and strategies of the enterprise evolve, it is natural for the risk environment to also evolve. Management should be kept aware by the risk practitioner of changes in risk over time and the impact of changes in the risk environment. Elimination of risk response is an infeasible goal. However, risk should be considered as part of strategic planning so that changes can be anticipated and mitigated as much as possible at the time that new strategic plans are unveiled, designed and implemented, rather than relying on the risk response process to address these retroactively.

3.10 Risk Treatment Plans

As previously discussed, there are several responses to risk that may be considered, and the risk practitioner plays a consultative role in assisting risk owners with deciding on the correct response to a risk: accept, transfer, mitigate or

CHAPTER 3— RISK RESPONSE AND REPORTING

avoid. The ultimate decision on risk response lies with the risk owner, but the risk practitioner can and should provide advice on technologies, policies, procedures, control effectiveness and leveraging of existing controls.

The decision plan to implement a particular risk treatment is based on many factors, including but not limited to the following:

- Current risk level
- Applicable laws and regulations
- Ongoing projects
- Strategic plans and management priorities
- Current and projected budgets
- Availability of staff
- Public pressure
- Actions of competitors

The risk response process shown in **figure 3.13** reflects the following concepts:

1. Risk scenarios (from risk identification) drive the risk analysis and assessment.
2. Risk analysis (from risk assessment) leads to the documentation (mapping) of risk.
3. Risk response is determined by the risk appetite of the enterprise. Risk within the appetite should be accepted. If the risk exceeds the appetite, then a risk response is needed.
4. The risk response options are considered along with parameters to determine the best available risk response.
5. The selected risk response is documented.
6. Risk responses are prioritized according to the current risk environment and cost-benefit analysis or another value-driven methodology.
7. A risk treatment plan is created to manage the risk response projects, reviewed and approved by the risk owner and added to the risk register.

CHAPTER 3— RISK RESPONSE AND REPORTING

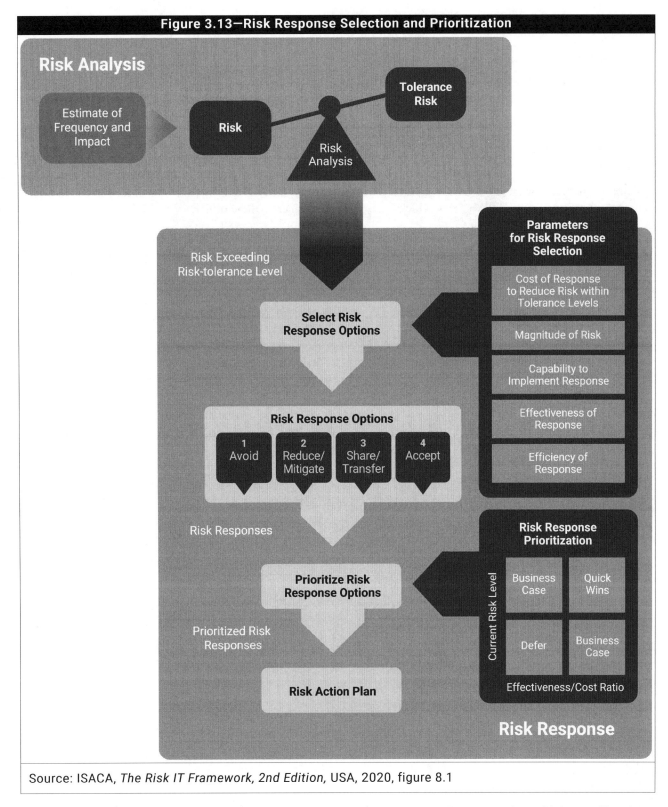

Figure 3.13—Risk Response Selection and Prioritization

Source: ISACA, *The Risk IT Framework, 2nd Edition*, USA, 2020, figure 8.1

Risk response is subject to the strategic direction and technological direction of the enterprise, which can affect the basic response selection (particularly the decision to transfer versus mitigate), the choice of one control over another and its schedule for deployment.

Where mitigation is the chosen response, decisions can be difficult to make and often require some method of comparing control options. Questions that the risk practitioner should keep in mind include:

- Would the control be effective?
- Would the control provide a satisfactory ROI (in enterprises where this is a driving concern)?
- Does the enterprise have sufficient skill to implement, configure and maintain the control?
- Is there sufficient budget and time to implement the control?
- How much would the control cost to operate annually, or what would the impact of the control be on productivity?

Risk treatment planning should be run as a project, with a defined start and end date. The end date is often used to determine the critical path of the project, which refers to those elements of the project that may have a direct impact on whether the end date can be met.

The risk practitioner should understand that changes in the delivery of any project element on the critical path affect the delivery of the entire project. For example, a project that does not receive its equipment from the supplier on time may not be able to meet the scheduled project dates. Critical path elements should be given special consideration because delays in these elements increase overall project risk. Through experience and careful evaluation, the risk practitioner can maintain close awareness of the critical path and regularly advise the risk owner on the feasibility of meeting the scheduled end date.

3.11 Data Collection, Aggregation, Analysis and Validation

The risk practitioner has a wealth of data sources available, including network devices, application logs threat intelligence and audit reports. Having data is valuable, but events may be hidden by sheer volume. Therefore, the incorrect analysis of data may lead to an erroneous conclusion.

The challenges of data analysis start with the completeness and trustworthiness of the data:

- Are all of the data available?
- Have any of the data been altered or changed?
- Are the data in the correct format?
- Are the data based on measuring important factors?

Some of the most common approaches to conducting data analysis are listed in **figure 3.14**.

| Figure 3.14—Methods for Uncovering Less Obvious Risk Factors ||
Method	Description
Cause-and-effect analysis	A predictive or diagnostic analytical tool that is used to: • Explore the root causes or factors that contribute to positive or negative effects or outcomes • Identify potential risk **Note:** A typical form is the Ishikawa or "fishbone" diagram.
Fault tree analysis	A technique that: • Provides a systematic description of the combination of possible occurrences in a system, which can result in an undesirable outcome (top-level event) • Combines hardware failures and human failures A fault tree is constructed by relating the sequences of events that, individually or in combination, could lead to the top-level event, then deducing the preconditions for the: • Top-level event • Next levels of events, until the basic causes are identified (elements of a "perfect storm," or unlikely simultaneous occurrence of multiple events that cause an extraordinary incident) **Note:** The most serious outcome is selected as the top-level event.
Sensitivity analysis	A quantitative risk analysis technique that: • Helps to determine which risk factors potentially have the most impact • Examines the extent to which the uncertainty of each element affects the target object when all other uncertain elements are held at baseline values **Note:** The typical display of results is in the form of a tornado diagram.

When analyzing data to determine risk levels, the risk practitioner should be attentive to the trends of events in the data sources. Trends may reveal an emerging threat, but detection of threats that are not already known is possible only when regular analysis has already established a reliable baseline of what constitutes normal behavior. With a baseline established, deviations more easily draw attention.

3.11.1 Data Collection and Extraction Tools and Techniques

The risk practitioner may use various sources of data to monitor and report on risk. Common internal data sources include:

- Prior risk assessments
- Project documents, especially risk logs and documented lessons learned
- Tickets from change, problem, release, configuration, asset and incident management systems
- Audit and incident reports
- User feedback and observation
- Interviews with management
- Security and test reports, including reports from tests of continuity and recovery plans
- Event and activity logs
- Independent third-party accredited assessment report, certification, etc.
- Internal control reviews conducted by the risk practitioner

Operational staff frequently aggregate data from multiple sources for broader visibility. Keep in mind that aggregated data typically present a summary view that may obscure details present in individual sources. Risk practitioners searching for unusual activity should keep in mind that detailed analysis of original sources may reveal patterns not identifiable under aggregation.

Data should also be validated to ensure its quality, including verifying that it is of the expected type, falls within plausible ranges, and has logical consistency across fields. For instance, data intended to present dates and times should not be text strings, and a delivery date should never be earlier than the order date. Where validation fails, the risk practitioner should investigate whether the problem arose from how the data was retrieved or is present in the source itself. The former typically suggests a procedural problem, while the latter can indicate misconfiguration or even be an indicator of compromise.

Logs

Logging is commonly provided by systems, devices and applications and remains the most consistently popular way to capture and store data for analysis. Analysis of log data can identify security violations and be instrumental in forensics investigations. Log analysis can alert the organization to malicious activity, such as a developing attack or multiple attempts to break in. Log data may be used to identify the source of an attack and assist in tailored strengthening of controls.

Logging has traditionally presented a trade-off between speed, detail and utility. If a log contains too much data from too many disparate sources, it may be difficult to notice significant individual events. Time synchronization of log entries can assist with correlation of events from multiple sources and improve the usefulness of review. Logging also takes time, potentially decreasing throughput for each transaction monitored. By understanding the ways in which log data is obtained, the risk practitioner is better able to anticipate the extent to which the data captured presents a complete picture of activity.

Analysis of log data and control activity should answer the following questions:

- Are the controls operating correctly?
- Is the level of risk acceptable?
- Are the controls aligned with the risk strategy, business strategy and key priorities?
- Are the controls flexible enough to meet changing threats?
- Are the correct risk data being provided in a timely manner?
- Is the risk-management effort benefiting corporate objectives (or at a minimum, not hindering them)?
- Is awareness of risk and compliance requirements reflected in user behavior?

As network and client security systems continue to mature, logging for specific purposes is becoming more granular and offering a greater degree of detail for analysis. For instance, client-resident data loss prevention (DLP) software can integrate with antivirus and anti-malware modules to distinguish between interactive (human) and process-driven attempts to migrate data in ways that are forbidden. Such information might reveal the difference between inadequate user training (or a malicious insider) and a remotely compromised system. Logging by intrusion detection or prevention systems (IDS/IPS) placed within the network can also be useful in detecting suspicious traffic patterns, particularly when combined with advanced behavior-based (heuristic) analysis. These capabilities are increasingly present in security baselines of cloud hosting environments and can leverage vastly more processing power than was possible just a few years ago to identify patterns that localized analysis would not be able to detect.

Logs may contain information that is sensitive or needed for forensic purposes, so they should be configured in ways that prevent alteration or deletion as well as preventing access by authorized personnel. In particular, SoD should be implemented in ways that prevent administrators with responsibility for systems or applications from altering or deleting logs made against their own scopes of responsibility. The risk practitioner should consider log access permissions as part of evaluating the level of potential insider threat.

CHAPTER 3— RISK RESPONSE AND REPORTING

Logs are one of the most valuable tools to monitor controls and detect risk, but they are often underutilized. Common challenges relating to the effective use of logs include:

- Having too much data
- Difficulty in searching for relevant information
- Improper configuration (e.g., may not be enabled or contain appropriate data)
- Modification or deletion of data before it is read (e.g., too little storage space)

An effective log should contain a record of all important events that occur on a system, such as:

- Changes to permissions
- System startup or shutdown
- Login or logout
- Changes to data
- Errors or violations
- Job failures
- Who (name/user ID) performed changes or modified the configuration
- Time when the event occurred

Not only can log reviews identify risk-relevant events such as compliance violations, suspicious behavior, errors, probes or scans and other abnormal activity, log review is often the most effective and sometimes the only feasible way of detecting certain indicators of compromise. Failure to review logs can result in the enterprise not being aware of an ongoing attack. Logs should also be preserved for forensic analysis if needed at a later time. The risk practitioner may find it useful to employ analysis tools to filter pertinent data in logs. Advanced algorithms also exist that can rapidly parse log data and highlight activity that may warrant greater concern.

Security Information and Event Management

As the numbers of network end points and routes expand, the risk practitioner may become overwhelmed by the sheer volume of data presented by logs and other individually-based alerting mechanisms. Security information and event management (SIEM) systems are integrated data correlation tools that address this problem by capturing data from multiple sources and analyzing the system, application and network activity reported in these data feeds for possible security events. SIEM systems can be used to detect attacks in progress by signature or behavior (heuristics) as well as identify compliance violations. **Figure 3.15** illustrates a notional SIEM architecture.

CHAPTER 3— RISK RESPONSE AND REPORTING

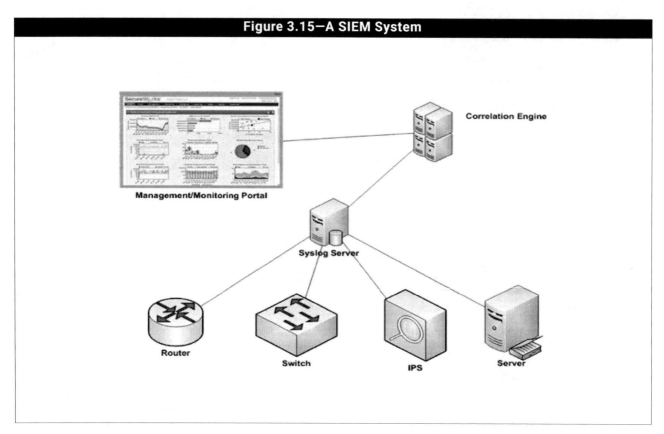
Figure 3.15—A SIEM System

The key to SIEM is the speed of the correlation engine. By gathering and correlating data from multiple sources, a SIEM system can develop reports on security at the enterprise level while retaining the ability to highlight relationships among activity on various parts of the network or systems. Correlation based on multiple criteria such as type, timing, chronological sequence and identified source of an event are all commonly available in SIEM implementations. In many cases, very granular assessment is possible, including activities associated with specific transaction codes, network IDs or types of traffic that may be generally permissible but are outside of common behavioral norms for the monitored network. SIEM is particularly effective in cloud environments where processing power can be provisioned on the fly to address particularly high analysis requirements in near-real time.

By highlighting developing trends, SIEM systems allow risk practitioners to identify risk and bring it to the attention of management before it would typically be discernible at the level of individual logs. Some large organizations may also maintain their own security operations centers (SOCs) that can perform additional correlation and analysis as well as near-real time monitoring.

Integrated Test Facilities

When seeking to monitor the performance and operation of an application, one common method is to use an integrated test facility (ITF). An ITF is a testing methodology that processes test data through production systems to test whether the systems are operating correctly. Within an ITF, the organization might set up several fictitious customers or transactions that are processed along with real data, allowing business analysts to observe the operation of the production systems and ensure correct processing. Of course, proper tracking of fictitious data is necessary to ensure that these test entries do not generate unintended downstream production activities.

CHAPTER 3— RISK RESPONSE AND REPORTING

External Sources of Information

Although internal sources provide the clearest picture of what is happening within the organization, the risk practitioner can gain additional insight by reviewing external data sources such as:

- Media reports
- Computer emergency (or incident) response team (CERT/CIRT) advisories
- Security company reports
- Regulatory bodies
- Peer organizations

Reports from antivirus and security companies, or from telecommunications providers such as the Verizon Data Breach Investigations report, are issued annually to provide a review of risk and exposure factors and may assist in developing and presenting a risk management solution to management. Information on incidents and risk may also come from computer or cybersecurity monitoring and reporting services by government and nonprofit organizations.

3.12 Risk and Control Monitoring Techniques

To support the ability to monitor and report on risk, the risk practitioner should ensure that processes, logs and audit hooks are commonly placed into the control framework, which allows for the monitoring and evaluation of controls. As controls are designed, implemented and operated, the risk practitioner should ensure that logs are enabled, controls are able to be tested and regular reporting procedures are developed.

The actual monitoring and reporting of controls should be performed in the risk monitoring phase of risk management; however, the risk practitioner should ensure that the capability to monitor a control and to support monitoring systems is addressed in control design. If the organization is using a managed security service provider (MSSP) or a SIEM system, the ability to capture data and the notification to the operations staff on the deployment of the system are necessary.

3.12.1 Monitoring Controls

The risk practitioner should work with stakeholders to set up an IS control monitoring process that reflects objectives, scopes and methods consistent with enterprise objectives and integrate this approach with performance management at the enterprise level.

In order to prevent an acceptance of risk within IT that exceeds the risk acceptance criteria set by the business, risk monitoring and evaluation should be integrated with enterprise performance management systems to ensure alignment between IT risk and business risk. The controls mandated through risk management must be aligned with IT security and related enterprise policies, subject to a regular process of review and revision to ensure that IT policies support the requirements of the enterprise.

The purpose of the IS control monitoring function is to ensure that IT security requirements are being met; standards are being followed; and staff is complying with the policies, practices and procedures of the enterprise.

Monitoring controls is a process that has six steps:

1. Identify and confirm risk control owners and stakeholders.
2. Engage with stakeholders and communicate the risk and information security requirements and objectives for monitoring and reporting.
3. Align and continually maintain the information security monitoring and evaluation approach with IT and

CHAPTER 3— RISK RESPONSE AND REPORTING

enterprise approaches.

4. Establish information security monitoring processes and procedures.
5. Determine life cycle management and change control processes for information security monitoring and reporting.
6. Request, prioritize and allocate resources for monitoring information security.

The risk practitioner should remember that the purpose of a control is to mitigate a risk, so the purpose of control monitoring is to verify whether the control is effectively addressing the risk, not to see whether the control itself is working. For example, a firewall that is configured in a way that does not regulate traffic between two networks may be ineffective even if the firewall is operational and processing traffic.

Risk monitoring and evaluation are processes performed to:

- Collect, validate and evaluate business, IT and process goals and metrics
- Monitor processes to ensure that they are performing in line with established performance metrics
- Provide reports that are systematic and timely

To accomplish these tasks, data related to risk management must be gathered from various sources in a timely and accurate manner. After the data has been validated for integrity, analysis can be performed against specific performance targets. Properly done, this process provides a succinct, all-around view of IT performance within the enterprise monitoring system.

The risk practitioner should continuously monitor, benchmark and improve the IT control environment and control framework to meet organizational objectives. When the results of the monitoring indicate an area of noncompliance or unacceptable performance, the risk practitioner should recommend mitigation activities such as implementation of new controls, adjustment or enforcement of existing controls, or changes to a business process to address the risk environment.

The monitoring of controls and the risk management framework may be done through self-assessment or independent assurance reviews. Risk treatment plans required to address a risk should be monitored to ensure appropriate risk management practices in alignment with risk appetite and tolerance. Assurance activities performed by independent entities always should be separate from the function, group or organization being monitored, and these entities should have the necessary skills and competence to perform assurance and adhere to codes of ethics and professional standards.

Any exceptions to risk monitoring and control activities should be reported promptly, followed up with analysis and addressed with properly prioritized corrective actions. As the risk environment changes and internal control systems are affected by changes in business and IT risk, gaps in the risk environment must be evaluated and changes made where necessary.

Sources of control monitoring information include:

- Network operations centers (NOC)
- Security operations centers (SOC)
- Command and control centers
- Continuous control monitoring
- Periodic testing
- Independent assessments

The risk practitioner should encourage each process owner to take ownership of control improvement through a continuing program of self-assessment to evaluate the completeness and effectiveness of management's control over processes, policies and contracts. In addition, explicit acknowledgment by management that all monitoring activities are functioning as designed should be recorded in the risk registry and updated on a regular basis.

3.12.2 Control Assessment Types

Effective control monitoring depends on the accuracy and completeness of the data provided for evaluation. The risk practitioner must ensure that the data received has been generated from the true and correct source. Data that can be retrieved directly by the risk practitioner are preferable to data provided by a third party. The risk practitioner should encourage local ownership of risk and control monitoring, which helps instill a risk culture in which local managers accept responsibility for risk and enables a faster detection of violations and security incidents.

Self-assessment

Controls are most easily assessed by those who regularly interact with them as part of their job functions. Through facilitated workshops, surveys or sampling methods, operational staff can determine the general effectiveness of controls on a recurring basis. In time, these self-assessments establish baselines, which can be used to highlight deviations that may indicate control weaknesses or failures. Self-assessment is included in methodologies promulgated by NIST as well as the COBIT and COSO frameworks.

IS Audit

Audit teams provide independent and objective review of the effectiveness and appropriateness of the control environment. Information provided by IS auditors can underline the need for control enhancement and bring risk to the attention of management along with strong supporting evidence. By working with audit teams, the risk practitioner can align the risk management program with the audit program and may be able to provide supporting data to the IS auditors. Recommendations provided by IS audit often require the attention of the risk practitioner through the updating of risk treatment plans, the risk register and enhancement of controls.

Mechanized or automated assessment by an independent application is an emerging form of IS audit for technical controls that has the potential to revolutionize the process. Under this approach, a dedicated, secure capability for technical evaluation operates in parallel to the production process, evaluating 100% of control actions in near-real time based on predefined criteria. Operating in a manner analogous to an intrusion detection system, automated assessment does not impede the flow of production traffic as a general rule and can generate alerts when control failures are detected.

Vulnerability Assessment

A vulnerability assessment is a methodical review of security to ensure that there are no predictable and unaddressed attack vectors (such as unnecessary open ports or services) that could be used to compromise an environment intentionally or unintentionally. The scope of an assessment may range from a single system or application to an entire facility or end-to-end business process. The purpose of the assessment is to deliver information to management that can be used both to understand the effectiveness of the risk management program and make decisions regarding treatment of identified vulnerability, such as deploying new controls.

A risk practitioner participating in a vulnerability assessment should have a working knowledge of both technical and nontechnical security controls in the environment and the overall architecture. For example, a review of a firewall may include the configuration, the change control process for managing the configuration, the process for monitoring the firewall logs, the system architecture to ensure the firewall cannot be bypassed, and the training and competence levels of the firewall administrators. There are many open source and commercial tools that can be used to automate or supplement certain aspects of the vulnerability assessment process. Additionally, many websites list known vulnerabilities with common applications, operating systems (OSs) and utilities.

Regular, rigorous vulnerability assessment is an important tool in the risk-management process, but assessment alone is not a comprehensive solution. Vulnerabilities identified by an assessment may be false positives or may be addressed by compensating controls, and even genuine vulnerabilities may be accepted by management if the only available mitigation strategies cost more than the potential impact of compromise. Also, even the most rigorous vulnerability assessment is limited to weaknesses that are either generally known or can be discerned by the assessor based on the information at hand.

Risk practitioners should be careful to avoid the trap of reliance upon automated tools, particularly when assessing vulnerability of IT systems. New methods of compromising systems are continuously being discovered, both intentionally and unintentionally, and scanning detects only those weaknesses whose use in compromising other systems has already been publicized. Automated tools are best viewed as a way to save time on certain areas of assessment so the practitioner can devote more attention to other areas.

Penetration Testing

A penetration test is a targeted attempt to break into an environment. As with a vulnerability assessment, the target of a penetration test may be a system, application, facility or end-to-end business process.

If the goal of a penetration test is to validate a vulnerability assessment, the tester is typically given the results of the assessment in order to focus on attempts to exploit the vulnerabilities that have already been identified. If the penetration tester is able to break in, then the vulnerability is real and must be mitigated; otherwise, there is a good chance that the vulnerability may be a false positive and not require mitigation. This validation-centric approach is typically called a white hat penetration test.

Penetration tests can also be undertaken against systems believed to be secure. In order to test the effectiveness of the security associated with such a system, the tester is generally given no information beyond a limited number of ground rules for the engagement and a definition of what constitutes a successful compromise of the system. This approach closely approximates the circumstances of someone attempting to break into a system maliciously and is typically called a black hat penetration test. A group on a long-term engagement attempting to penetrate security mechanisms using black-hat methods, whether external to the organization or established from its own staff, is commonly called a Red Team. This designation contrasts with the Blue Team, the proactive cybersecurity defenders of the organization who confront both Red Team and real-world adversaries.[5]

In carrying out his or her work, a penetration tester often uses the same tools used by malicious hackers to try to break into systems. This approach provides meaningful results, but methods used for penetration testing may pose a risk to the enterprise in the form of system failure or data compromise. Therefore, it is of utmost importance that such tests—whether white hat or black hat—are conducted only with management approval, using a defined methodology and under proper oversight. It is also essential that the results of penetration testing be used as the basis for improvement, which is what distinguishes a penetration test from a practical audit. In environments where a Red Team operates, this improvement can be made nearly continuous through active engagement between the Red and Blue teams to review lessons learned—a partnership sometimes called the Purple Team to denote the commitment from both teams.

---

[5] All Blue Team members are cybersecurity defenders, but not all defenders are part of the Blue Team. This designation is specific to *proactive* defense.

Third-party Assurance

An enterprise can earn customer and shareholder confidence by having a third party provide assurance or attestation of the validity of the information security program. Third-party attestation may be based on an external IS audit or a certification of compliance with an internationally recognized standard, such as COBIT 2019 or *ISO/IEC 27001—Information security management* or an industry standard, such as the PCI DSS. The third party is responsible for evaluating the processes of the subject enterprise and validating compliance with the requirements of the standard.

Another source for attestation is based on the Statement on Standards for Attestation Engagements Number 18 (SSAE 18), developed by the American Institute of Certified Public Accountants (AICPA). SSAE 18 standardizes System and Organization Controls (SOC) reports across three levels intended for different audiences, with varying levels of information disclosure based on sensitivity. This form of third-party assurance is used by many enterprises when relying on cloud or third-party service suppliers, where regular or continuous monitoring is particularly important to establish stakeholder confidence in the services being delivered outside the control of the organization but an individual right to audit is impractical.

3.13 Risk and Control Reporting Techniques

Risk practitioners are commonly expected to provide regular reports to management and the board of directors on the status of the risk management program and the overall risk profile of the enterprise. Making such a report requires the review of the effectiveness of the controls in the enterprise and their compliance with established policy. Controls may need adjustment, replacement or removal depending on the changes in the risk environment and the acceptance and appropriateness of the controls.

The effectiveness of control monitoring is dependent on the following:

- Timeliness of the reporting—are data received in time to take corrective action?
- Skill of the data analyst—does the analyst have the skills to properly evaluate the controls?
- Quality of monitoring data available—are the monitoring data accurate and complete?
- Quantity of data to be analyzed—can the risk practitioner find the important data in the midst of all the other log data available?

In addition, control owners may expect reports specific to their areas of accountability.

The channels best suited to reporting at different levels will depend on preferences of those receiving the information and may vary from one risk owner to another. The essential element of the risk practitioner's internal risk reporting obligations is that the information presented be clear, useful and timely, not that it adhere to a particular format. However, certain basic formats are commonly used in reporting risk, including heat maps, scorecards and dashboards.

3.13.1 Heat Maps

The goal of a heat map is to create a graphical representation of risk distribution across a particular scope, which might be a department, system or enterprise. The map itself is a scatterplot of data points against an X axis representing frequency and a Y axis representing impact. Only positive values are shown, so that the bottom left corner of the map is the origin (0,0). The plot is divided into squares, which may be 2×2, 3×3, 4×4, etc. Of these, the bottom left square is typically shaded green, while the upper right square is red. Between these extremes, squares are shaded in intermediate colors, so that it is apparent at a glance which (or how many) risk elements are at higher levels of impact, frequency or both.

Visually attractive, heat maps are widely regarded as having minimal informational value when based on qualitative assessment methods, which tend to defy meaningful aggregation. However, they can be useful as a means of expressing the effectiveness of controls within an overall architecture, localizing rates of compromise to particular systems or departments or presenting quantitative results. The essential element for a heat map is that it makes clear what it is presenting in terms of its underlying data, so that viewers can understand any practical limitations on the conclusions that can be drawn from it.

Figure 3.16 shows an example of a heatmap.

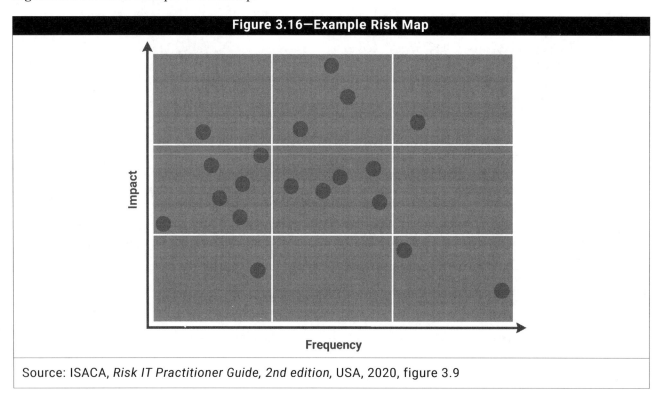

Source: ISACA, *Risk IT Practitioner Guide, 2nd edition*, USA, 2020, figure 3.9

3.13.2 Scorecards

Similar to academic grade reports, risk scorecards seek to simplify risk reporting by aggregating performance across particular functional areas and assigning grades or scores to each area. Scoring is typically done in a manner that reflects alignment with the enterprise risk appetite, although other bases are possible.

The original model for a risk scorecard suggested keeping risk and performance in distinct channels of reporting. Modern business or enterprise scorecards tend to incorporate risk indicators alongside the four main areas of the classic business balanced scorecard: financial, internal business processes, learning and innovation and clients and stakeholders. This approach unifies considerations of risk and performance against the backdrop of the overall strategic business context.

As with heat maps, scorecards are susceptible to biases in data arising from the limitations of qualitative assessments, whose results are not easily aggregated. However, this limitation can be mitigated by having a rigorous and effective process for identifying key risk indicators.

Figure 3.17 shows the dimensions of the classic balanced scorecard, where risk indicators might be added alongside performance indicators as part of the measurements for each area.

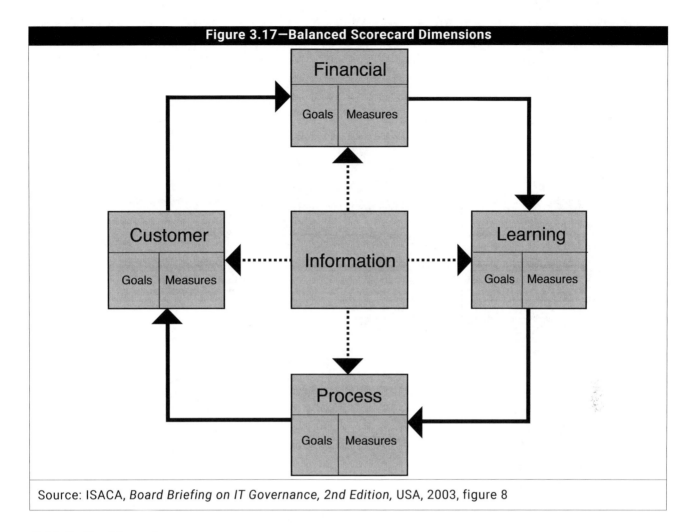

Figure 3.17—Balanced Scorecard Dimensions

Source: ISACA, *Board Briefing on IT Governance, 2nd Edition*, USA, 2003, figure 8

3.13.3 Dashboards

When data are presented sequentially with distinct indicators for each item, the presentation is commonly called a dashboard. Values assigned to each dashboard item may be numeric scores, color-coded "stoplight" indicators or a combination of the two. The intent is to allow rapid focus on those items that are falling outside acceptable levels. In practice, the effect of a dashboard tends to be elimination of focus from anything that fails to draw sufficient attention.

As with a heat map, a dashboard is limited by the perception that aggregate risk adheres to particular statuses intended for isolated analysis. This perception is accurate for qualitative values but unreliable when assessments are qualitative in nature, as many tend to be. The result is a breakdown in which two "yellow" items may be vastly different but receive the same attention. Risk practitioners should take care to understand the underlying data used to create stoplight charts and other forms of risk dashboard, so that senior managers reviewing these presentations can be made aware of their limitations and implications.

Metrics reported on dashboards or by similar means should be measured consistently on a recurring basis (e.g. quarterly) to facilitate trend identification and analysis. Remediation or response actions associated with particular metrics should also be clearly documented.

Figure 3.18 shows an example of a typical risk dashboard.

CHAPTER 3— RISK RESPONSE AND REPORTING

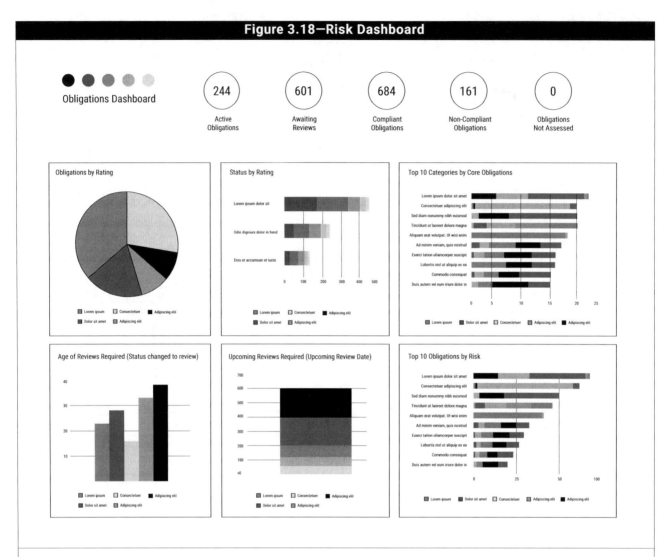

Figure 3.18—Risk Dashboard

Source: InetSoft, "OEM Case Study: Take Control of Your Enterprise Governance, Risk and Compliance," https://www.inetsoft.com/solutions/risk_management_reporting/. Source image courtesy of Protecht.

3.14 Key Performance Indicators

Performance indicators measure how well a process is performing in terms of its stated goal. This measurement provides insight into whether action may be required, with the goal of allowing changes to be made before significant impacts occur. Although many indicators loosely correlate with performance, a subset of these indicators are effective in predicting whether organizational goals will be reached and indicates the capabilities, practices and skills of value to the organization. These closely correlated performance indicators are called KPIs.

KPIs measure activity goals, which are actions that process owners must take to achieve effective process performance. They are commonly used to set benchmarks for risk management goals and to monitor whether those goals are being attained. Management sets its risk management goals according to its risk acceptance level and desired cost-benefit analysis. For example, a KPI may indicate that an error rate of five percent is acceptable, which implies that an error rate higher than five percent is unacceptable and requires escalation with some form of response.

A KPI should be based on SMART metrics:

- **Specific**—based on a clearly understood goal; clear and concise
- **Measurable**—able to be measured; quantifiable (objective), not subjective
- **Attainable**—realistic; based on important goals and values
- **Relevant**—directly related to a specific activity or goal
- **Timely**—grounded in a specific time frame

Additionally, KPIs should be:

- Valuable to the business
- Tied to a business function or service
- Under the control of management
- Quantitatively measured
- Usable in different reporting periods (for consistency)

Examples of potential KPIs include:

- Network availability
- Customer satisfaction
- Number of complaints resolved on first contact
- Time between data request and presentation
- Number of employees that attended awareness sessions

KPIs place emphasis on processes that should be good indicators of the health of the overall process. They are often used on charts or graphs to report compliance to management in a clear, easily understood manner and have long been included in balanced scorecard reporting.

3.15 Key Risk Indicators

Risk indicators are used to measure risk levels in comparison to defined risk thresholds, so that the organization receives an alert when a risk level approaches an unacceptable level. By putting tracking and reporting mechanisms in place that alert staff to a developing or potential risk, the enterprise gains the opportunity to respond to the risk before it produces unacceptable outcomes.

Key risk indicators (KRIs) comprise a subset of risk indicators that are highly relevant and possess a high probability of predicting or indicating important risk. Examples of KRIs include:

- Number of unauthorized equipment or software detected in scans
- Number of instances of SLAs exceeding thresholds
- High average downtime due to operational incidents
- Average time to deploy new security patches to servers
- Excessive average time to research and remediate operations incidents
- Number of desktops/laptops that do not have current antivirus signatures or have not run a full scan within scheduled periods

KRIs support numerous aspects of risk management, including:

- Risk appetite, by validating the enterprise's risk appetite and risk tolerance levels
- Risk identification, by providing an objective means for identifying risk
- Risk mitigation, by providing a trigger for investigating an event or providing corrective action

CHAPTER 3— RISK RESPONSE AND REPORTING

- Risk culture, by helping the enterprise focus on important, relevant areas
- Risk measurement and reporting, by providing objective and quantitative risk information
- Regulatory compliance, by providing data that can be used as an input for operation risk capital calculations

3.15.1 KRI Selection

KRIs should be selected carefully and sparingly. Common mistakes made when implementing KRIs include regarding too many risk indicators as being KRIs and choosing KRIs that are flawed in some way. These include those KRIs that:

- Are not linked to specific risk
- Are incomplete or inaccurate due to unclear specifications
- Are difficult to measure, aggregate, compare and interpret
- Provide results that cannot be compared over time
- Are not linked to goals

The effectiveness of KRIs depends in large part on the strength of their metrics. As with KPIs, KRIs should leverage SMART metrics. Other factors that can influence the selection of KRIs include:

- **Balance**—the set of selected risk indicators should include:
  - Lagging indicators (indicating risk after events have occurred)
  - Leading indicators (indicating which controls are in place to prevent events from occurring)
  - Trends (analyzing indicators over time or correlating indicators to gain insights)
- **Root cause**—selected indicators should drill down to the root cause of events, not just the symptoms.

The selection of an appropriate set of KRIs benefits the organization by:

- Providing an early warning (forward-looking) signal that a high risk is emerging to enable management to take proactive action (before the risk actually becomes a loss)
- Providing a backward-looking view on risk events that have occurred, enabling risk responses and management to be improved
- Enabling the documentation and analysis of trends
- Providing an indication of the enterprise's risk appetite and tolerance through metric setting (i.e., KRI thresholds)
- Increasing the likelihood of achieving the enterprise's strategic objectives
- Assisting in continually optimizing the risk governance and management environment

When working with management to determine appropriate KRIs, the risk practitioner should work with all relevant stakeholders to ensure greater buy-in and ownership. Risk indicators should be identified for all stakeholders, and IT-based metrics should be aligned with other metrics used in the enterprise to the greatest extent possible.

3.15.2 KRI Effectiveness

KRI effectiveness takes into consideration the following criteria:

- **Impact**—indicates risk with high business impact.
- **Effort**—is the easiest to measure and maintain among indicators of equivalent sensitivity.
- **Reliability**—possesses a high correlation with the risk and is a good predictor or outcome measure.

- **Sensitivity**—is capable of accurately indicating risk variances.
- **Repeatable**—can be measured on a regular basis to show trends and patterns in activity and results.

3.15.3 KRI Optimization

To ensure accurate and meaningful reporting, KRIs should be optimized to ensure that the correct data are being collected and reported on and thresholds are set correctly. KRIs that report on data points that either cannot be controlled by the enterprise or are not alerting management to an adverse condition at the correct time should be adjusted to be more precise, relevant or accurate.

Figure 3.19 describes examples of KRI optimization.

Figure 3.19—Examples of KRIs Optimized for Management	
Metric Criterion	Description
Sensitivity	Management has implemented an automated tool to analyze and report on access control logs based on severity, and the tool generates an excessive number of results. Management performs a risk assessment and decides to configure the monitoring tool to report only on alerts marked "critical."
Timing	Management has implemented strong segregation of duties (SoD) within the enterprise resource planning (ERP) system. One monitoring process tracks system transactions that violate the defined SoD rules before month-end processing is completed so that suspicious transactions can be investigated before reconciliation reports are generated.
Frequency	Management has implemented a key control that is performed multiple times a day. Based on a risk assessment, management decides that the monitoring activity can be performed weekly because this will capture a control failure in sufficient time for remediation.
Corrective action	Management has implemented a remediation process to bring controls into alignment with the organizational risk appetite. Using existing problem management tools, management is able to integrate automated monitoring of the controls in the process to prioritize existing gaps, assign problem owners and track remediation efforts.

3.15.4 KRI Maintenance

Because the organization's internal and external environments are constantly changing, the risk environment is also highly dynamic. The set of KRIs should be evaluated on a regular basis to verify that each indicator remains properly related to the risk appetite and tolerance levels of the enterprise, and that trigger levels are defined at points that allow stakeholders to take appropriate action in a timely manner. Any KRIs that are no longer related to the risk appetite and tolerance should be replaced, while any whose trigger levels are found to be out of alignment with the requirements of the enterprise should be optimized.

3.15.5 Using KPIs with KRIs

KPIs and KRIs are often used in conjunction with one another to measure performance and mitigate risk. KPIs help to identify underperforming aspects of organizations and areas of the business that may require additional resources and attention, while KRIs provide early warnings of increased risk within the enterprise. Failing to meet performance goals is itself a risk, so the two can be used together in risk management or combined with indicators specific to controls that may foreshadow an increase in risk due to a decrease in the effectiveness of mitigation.

To illustrate the difference, consider the risk that unpatched systems may lead to a serious breach of customer data, loss of availability and financial loss. In response to this risk, the organization develops a policy to apply all critical patches within 30 days. The development of this policy is an example of a KPI: the enterprise has a clear, measurable standard of what constitutes proper performance. On a scheduled basis, the enterprise tracks and reports to senior management on whether that KPI has been met and reports on the average time to deploy critical security patches.

If an organization tracks the time between the release of a patch and its deployment within the organization on a monthly basis, then a report that shows the average time taken could be developed, as shown in **figure 3.20**.

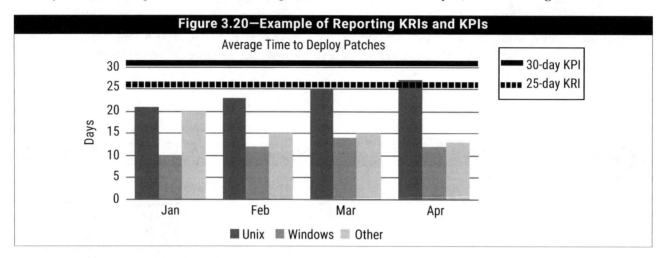

Figure 3.20—Example of Reporting KRIs and KPIs

Through this chart, a manager can see that the time to deploy patches on a UNIX system is increasing each month and is in danger of exceeding the 30-day threshold. To detect a trend or developing problem, the organization may also set a KRI that will represent a threshold or level of performance that is getting dangerously close to exceeding the KPI but has not yet done so. By setting a KRI of 25 days, management can see that the time to deploy a UNIX patch has now increased to the point of triggering an alert. This alert should precipitate a review of the UNIX patching procedures, and management should investigate the situation to enable preventive action and address the slow deployment process before the deployments of these patches violates the policy.

This example shows how KRIs and KPIs can be used to monitor and track the attainment of goals. When measuring the success of a risk management program, KPIs must be based on meaningful criteria that can help management track the overall success of the controls put in place to support risk response.

3.16 Key Control Indicators

Certain indicators reveal the effectiveness of controls. Of these, the subset that quantifies how well a specific control is working constitutes the set of key control indicators (KCIs). The goal of KCIs is to track performance of control actions relative to tolerances, providing insight into the ongoing adequacy of a given control in keeping risk within acceptable levels. For this reason, they are sometimes called control effectiveness indicators.

Examples of KCIs include the numbers of:

- Phishing emails not blocked by filtering systems (indicating a weakness in features or tuning and foreshadowing a higher risk of compromise linked to phishing)
- Administrators with rights to both test and production environments (indicating a weakness in procedures intended to establish and maintain segregation of duties)
- Unsolicited prompts for second-factor authentication (indicating a weakness in password security allowing would-be hackers to reach the second stage of authentication)

Each KCI has its own tolerances based on the underlying assets that its controls protect and the impact to the organization associated with their loss or compromise. Because controls affect risk, KCIs are correlated not only with specific controls but also with the risk that those controls are meant to mitigate. As a result, KCIs tend to be

CHAPTER 3— RISK RESPONSE AND REPORTING

leading indicators for KRIs associated with the controls that they monitor: where tolerances are broken, the KCI reveals an increasing risk exposure that will eventually be represented in the KRIs associated with the monitored control.

This correlation also means that all KCIs are themselves risk indicators; however, they rarely rise to the level of being KRIs in their own right, tending instead to be secondary indicators of risk arising from control failure or nonperformance. Additionally, because individual controls may affect multiple risk areas, KCIs may have broader implications than KRIs.

Chapter 4:
Information Technology and Security

Overview

Domain 4 Exam Content Outline 190
Learning Objectives/Task Statements 190
Suggested Resources for Further Study 191

Part A: Information Technology Principles

4.1 Enterprise Architecture 195
4.2 IT Operations Management..197
4.3 Project Management 216
4.4 Enterprise Resiliency 218
4.5 Data Life Cycle Management 221
4.6 System Development Life Cycle..224
4.7 Emerging Trends in Technology 226

Part B: Information Security Principles

4.8   Information Security Concepts, Frameworks and Standards...234
4.9   Information Security Awareness Training 246
4.10 Data Privacy and Principles of Data Protection 247

CHAPTER 4— INFORMATION TECHNOLOGY AND SECURITY

Overview

IT risk management is distinct from IT operations or information security; however, it cannot be fully separated from these disciplines. To be effective in the management of IT risk, the risk practitioner must have a working understanding of key principles and concepts of information technology and security. This chapter aims to provide that foundational knowledge, including familiarity with computer hardware, software and networking concepts; enterprise resiliency; secure system development; the importance of training and awareness in the modern workforce; and familiarity with data privacy.

This chapter represents 22 percent (33 questions) on the CRISC exam.

Domain 4 Exam Content Outline

A. Information Technology Principles

1. Enterprise Architecture
2. IT Operations Management
3. Project Management
4. Disaster Recovery Management
5. Data Life Cycle Management
6. System Development Life Cycle
7. Emerging Technologies

B. Information Security Principles

1. Information Security Concepts, Frameworks and Standards
2. Information Security Awareness Training
3. Business Continuity Management
4. Data Privacy and Data Protection Principles

Learning Objectives/Task Statements

Upon completion of this chapter, the risk practitioner will be able to:

1. Collect and review existing information regarding the organization's business and IT environments.
2. Identify potential or realized impacts of IT risk to the organization's business objectives and operations.
3. Identify threats and vulnerabilities to the organization's people, processes and technology.
4. Evaluate threats, vulnerabilities and risk to identify IT risk scenarios.
5. Establish accountability by assigning and validating appropriate levels of risk and control ownership.
6. Facilitate the identification of risk appetite and risk tolerance by key stakeholders.
7. Promote a risk-aware culture by contributing to the development and implementation of security awareness training.
8. Review the results of risk analysis and control analysis to assess any gaps between current and desired states of the IT risk environment.
9. Collaborate with risk owners on the development of risk treatment plans.
10. Collaborate with control owners on the selection, design, implementation and maintenance of controls.

CHAPTER 4— INFORMATION TECHNOLOGY AND SECURITY

11. Evaluate emerging technologies and changes to the environment for threats, vulnerabilities and opportunities.
12. Evaluate alignment of business practices with risk management and information security frameworks and standards.

Suggested Resources for Further Study

Breaux, Travis; *An Introduction to Privacy for Technology Professionals*, International Association of Privacy Professionals, USA, 2014

ISACA, *Cybersecurity Fundamentals Study Guide*, 3rd 2dition, USA, 2021

ISACA, *ISACA Privacy Principals and Program Management Guide*, USA, 2016

SELF-ASSESSMENT QUESTIONS

CRISC self-assessment questions support the content in this manual and provide an understanding of the type and structure of questions that have typically appeared on the exam. Questions are written in a multiple-choice format and designed for one best answer. Each question has a stem (question) and four options (answer choices). The stem may be written in the form of a question or an incomplete statement. In some instances, a scenario or a description problem may also be included. These questions normally include a description of a situation and require the candidate to answer two or more questions based on the information provided. Many times, a question will require the candidate to choose the **MOST** likely or **BEST** answer among the options provided.

In each case, the candidate must read the question carefully, eliminate known incorrect answers and then make the best choice possible. Knowing the format in which questions are asked, and how to study and gain knowledge of what will be tested, will help the candidate correctly answer the questions.

1. Which of the following business requirements **BEST** relates to the need for resilient business and information systems processes?

    A. Effectiveness
    B. Confidentiality
    C. Integrity
    D. Availability

2. An information system that processes weather forecasts for public consumption is **MOST** likely to place its highest priority on:

    A. nonrepudiation.
    B. confidentiality.
    C. integrity.
    D. availability.

3. The **BEST** control to prevent unauthorized access to an enterprise's information is user:

    A. accountability.
    B. authentication.
    C. identification.
    D. access rules.

CHAPTER 4— INFORMATION TECHNOLOGY AND SECURITY

4. Which of the following controls **BEST** protects an enterprise from unauthorized individuals gaining access to sensitive information?

   A. Using a challenge response system
   B. Forcing periodic password changes
   C. Monitoring and recording unsuccessful logon attempts
   D. Providing access on a need-to-know basis

5. Which of the following defenses is **BEST** to use against phishing attacks?

   A. An intrusion detection system
   B. Spam filters
   C. End-user awareness
   D. Application hardening

Answers on page 193

CHAPTER 4— INFORMATION TECHNOLOGY AND SECURITY

Chapter 4 Answer Key

Self-assessment Questions

1. A. Effectiveness deals with information being relevant and pertinent to the business process and being delivered in a timely, correct, consistent and usable manner. While the lack of system resilience can in some cases affect effectiveness, resilience is more closely linked to the business information requirement of availability.
   B. Confidentiality deals with the protection of sensitive information from unauthorized disclosure. While the lack of system resilience can affect data confidentiality, resilience is more closely linked to the business information requirement of availability.
   C. Integrity relates to the accuracy and completeness of information and to its validity in accordance with business values and expectations. While the lack of system resilience can in some cases affect data integrity, resilience is more closely linked to the business information requirement of availability.
   D. Availability relates to information being available when required by the business process—now and in the future. Resilience is the ability to provide and maintain an acceptable level of service during disasters or when facing operational challenges.

2. A. Nonrepudiation refers to the ability to verifiably prove the originator of data, which is unlikely to be of importance for weather forecasts that are rendered accurately.
   B. Keeping data confidential would be at odds with the business purpose of a system designed to provide data for public use.
   C. A system that delivers weather forecasts is likely to place its highest priority on the integrity of the data. The risk practitioner should keep in mind that whether a forecast turns out to be accurate in its prediction is distinct from whether the data was accurately represented.
   D. Availability of data is likely to be a lower priority for a weather-forecasting system than the accuracy with which the data are presented.

3. A. User accountability does not prevent unauthorized access; it maps a given activity or event back to the responsible party.
   B. Authentication verifies the user's identity and the right to access information according to the access rules.
   C. User identification without authentication does not grant access.
   D. Access rules with the appropriate identification and authentication methods prevent unauthorized access.

4. A. Verifying the user's identification through a challenge response does not completely address the issue of access risk if access was not appropriately designed in the first place.
   B. Forcing users to change their passwords does not guarantee that access control is appropriately assigned.
   C. Monitoring unsuccessful access logon attempts does not address the risk of appropriate access rights.
   D. Physical or logical system access should be assigned on a need-to-know basis (legitimate business requirements) and in ways that incorporate least privilege and segregation of duties (SoD).

5. A. An intrusion detection system (IDS) does not protect against phishing attacks because phishing attacks usually do not have the same patterns or unique signatures.
   B. While certain highly specialized spam filters can reduce the number of phishing emails that reach their addressees' inboxes, they are not as effective in addressing phishing attacks as end-user awareness.

C. Phishing attacks are a type of social engineering attack and are best defended by end-user awareness training.

D. Application hardening does not protect against phishing attacks because phishing attacks generally use email as the attack vector, with the end user, not the application, as the vulnerable point.

Part A: Information Technology Principles

IT has evolved from a support service to a business enabler whose loss would effectively cripple operations. This part introduces the risk practitioner to a series of concepts and principles that are widely known within IT circles and have implications for the management of information risk.

4.1 Enterprise Architecture

Strategic management of enterprise information technology begins with an enterprise-level understanding of the network and information architecture. Enterprise architecture (EA) delivers a view of the current state of IT, establishes a vision for a future state and generates a strategy to move from current to future conditions that minimizes business disruption. The enterprise view of IT demonstrates links between IT and organizational objectives and produces a view of current risk and controls to answer or enable answering four basic questions:

- Are we doing the right things?
- Are we doing them the right way?
- Are we getting them done well?
- Are we seeing expected benefits?

These "Four AREs," shown in **Figure 4.1**, comprise a cycle of examination, evaluation and adjustment. The questions and their answers serve as important continuous feedback to implementation of enterprise IT strategy under an approved architecture and ensure that IT projects deliver their intended value. An EA typically includes business functions or capabilities, human roles, the physical structure of the organization, key business applications, data stores, data flows, platforms, hardware and network or communications infrastructure. In short, an EA frames how information enables the organization to do whatever it does.

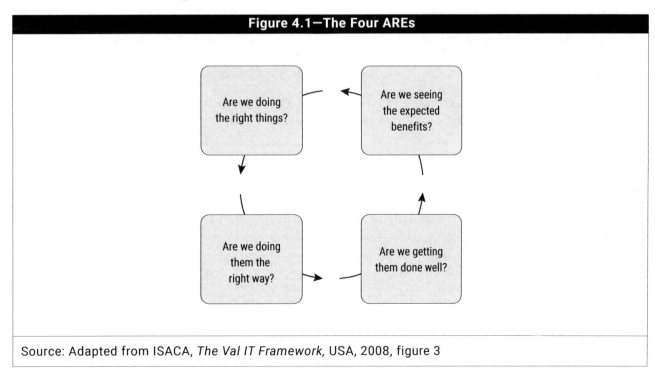

Source: Adapted from ISACA, *The Val IT Framework*, USA, 2008, figure 3

Although an enterprise architecture can be developed independently of any reference material, organizations commonly leverage EA frameworks to facilitate the process. All EA frameworks provide structured guidance across four key topics:

- **Documentation**—ways to present the architecture in terms of particular taxonomy or nomenclature depending on the audience
- **Notation**—ways to visualize the architecture in a standardized way
- **Process**—the goals, inputs, actions, and outputs that occur in building architecture, commonly grouped upwards into phases and supported from below by techniques and practices
- **Organization**—skill sets, training, and approaches to governance needed for the architecture

Different EAs address these topics differently. Architecture frameworks commonly used in commercial organizations include The Open Group Architecture Framework (TOGAF) and the Zachman Framework, named for John Zachman who pioneered the concept of EA in the 1980s. In the United States, the Department of Defense Architecture Framework (DODAF) is the standard for military enterprises, while the Federal Enterprise Architecture Framework (FEAF) is the standard for civilian agencies. Globally, the Sherwood Applied Business Security Architecture (SABSA) has gained prominence, and other EAs exist for specialized purposes such as telecommunications.

Risk management requires a complete and thorough assessment of the risk to each element of a system. The risk associated with an IT system is a composite of the risk associated with each element that makes up the system. Accordingly, risk in any one location can affect the security of all other areas, and a vulnerability on one system may affect the security of all other systems to which it connects.

In environments in which an EA is either immature or absent, the risk practitioner is effectively forced to place greater emphasis on the use of technology-specific assessments as means of building a piecemeal vision of current IT risk. Evaluation of risk against an established EA, therefore, yields both greater efficiency and more complete understanding of the risk environment.

4.1.1 Maturity Models

Since the 1980s, enterprises have looked for ways to baseline their standing in specific areas using survey-based frameworks. The goal of these maturity models is to not dictate how something should be done but rather what outcomes are desirable. By reviewing what they are doing relative to published baselines, organizations can gauge their progress in building mature programs as envisioned by leading enterprises and industry peers.

Maturity models designed specifically for EA have been published by numerous organizations over time, including NASCIO (EAMM) and Gartner (ITScore). The Open Web Application Security Project has published a Software Assurance Maturity Model (OWASP SAMM), and the NIST CSF (see section 3.6.1 Control Standards and Frameworks) has been used as a maturity model for cybersecurity. While specific models vary, they generally begin at a level where the program is either nonexistent or entirely notional, lacking any practical implementation. From there, organizations move through levels that include reaction, operation as distinct functions and integration as part of overall enterprise operations. This overall approach is consistent with principles of total quality management (TQM) and continuous process improvement (CPI), where the goal is to reach an optimizing or efficient level of operations through iteration and refinement.

In considering enterprise risk, the risk practitioner should inquire as to the existence of an EA and, where possible, assess the EA to determine its maturity. Significant effort, resources and time are required to develop an EA, and it is not uncommon for the risk practitioner to discover that the EA either has not been formally established or is at a low maturity state.

See chapter 3 Risk Response and Reporting for more information.

CHAPTER 4— INFORMATION TECHNOLOGY AND SECURITY

Note: Risk practitioners should be aware that some organizations are trending away from maturity models with regards to cybersecurity. TQM and CPI originated in factory production, where physical processes tended to limit the degree of monitoring that might be achieved. No such limits arise organically in information-centric organizations. Originating from the assumption that greater monitoring means less risk, maturity models tend to promulgate controls at increasing rates. The potential exists to expand monitoring to such a degree that the results of monitoring exceed an organization's ability to act upon them, yet the nature of cybersecurity risk is distinct from quality problems in manufacturing. A highly monitored environment may nonetheless fail to adequately manage risk.

The alternative to CMM is an approach that targets attention on measurements and controls that align with an organization's greatest known areas of risk. Organizations pursuing this approach believe that sustained commitments to cybersecurity, such as Red Team engagement (see section 3.12.2 Control Assessment Types), can lead to a proactive posture where security is truly part of daily operations. At present, there is no consensus around whether this dynamic represents an improvement over CMM, or instead defines "maturity" according to different criteria than those applied in the prevailing generation of models.

4.2 IT Operations Management

The risk practitioner is not expected or required to be a technical expert in the design, implementation or support of individual IT systems or applications. However, strong knowledge of general IT concepts is invaluable for anyone whose roles and responsibilities require close and ongoing interaction with IT staff. The identification, assessment and monitoring of IT risk, along with recommendation of appropriate responses to the risk that is identified, require that type of close and ongoing interaction. Accordingly, the risk practitioner should have a working knowledge of networking, applications and other aspects of IT that appear within the organizational IT infrastructure.

Particular areas of concern are identified for the risk practitioner in each of the following sections. These areas may encompass or substantially relate to particular or commonly identified threats, vulnerabilities or business impacts and are presented in the context of the risk-management process.

4.2.1 Hardware

IT systems and business processes rely on hardware—the equipment and devices that process, store and transmit data. Hardware includes many devices, such as:

- Desktop, laptop, or mobile computers, which include:
    - Central processing units (CPUs)
    - Motherboards
    - Random access memory (RAM)
    - Read-only memory (ROM)
- Networking components (switches, routers, etc.)
- Security components (firewalls, intrusion detection systems, etc.)
- Peripherals (keyboards, monitors, etc.)
- Industrial equipment enabled for networking ("smart" appliances/Internet of Things devices)

Areas of concern associated with hardware include:

- Obsolescence
- Poor maintenance
- Misconfiguration

CHAPTER 4— INFORMATION TECHNOLOGY AND SECURITY

- Use of default credentials
- Lack of secure configurations
- Unnecessary services, ports and protocols running on the system
- Missing or inaccurate documentation
- Physical loss or theft
- Data loss due to insecure disposal
- Unauthorized capture of traffic ("sniffing")
- Physical access
- Component or system failure
- Use without authorization
- Inaccurate asset inventory

Risk practitioners should keep in mind that "mobile computers" refers not only to tablets or two-in-one devices but also smartphones, which have evolved to the point of being full-featured systems for information gathering, storage, processing and transmission. In environments where users are permitted to have their personal mobile phones with them for use throughout the day, these devices may serve as unwitting channels by which data can be exfiltrated if they are compromised.

Additionally, on account of their small size, frequent use and the potential for quick resale on secondary markets at relatively high prices, mobile phones are among the most easily damaged, stolen and lost items in individual possession. Business processes that expect or require staff to use their personal mobile phones for communication may be affected by impacts to these devices, even though they are outside the scope of organizational ownership. This potential for business impact arising from personal loss is especially serious under paradigms that use personal mobile devices for tokens in multi-factor authentication (see section 4.8.5 Access Control).

Supply Chain Management

In recent years, there have been several examples of hardware intentionally infected with security vulnerabilities during the manufacturing or delivery process. These breaches have been found on network devices, point-of-sale terminals, applications and smartphones from numerous countries and vendors. However, such vulnerabilities are not easy to detect.

Risk practitioners should be aware of the risk of purchasing infected equipment and encourage their organizations to use trusted vendors or suppliers whenever possible. Purchasing equipment that has been tested and evaluated by an external entity using an internationally approved process, such as the Common Criteria (ISO/IEC 15408), may provide a higher level of confidence that the equipment is secure. Internal IT staff should also validate that maintenance hooks that allow vendors to gain access to systems for monitoring purposes are documented and either secured or eliminated. When performing administration of hardware devices, a secure channel should be used with strong authentication required.

4.2.2 Software

Software refers to the programming used to cause hardware to carry out specific functions or activities in ways that can be reconfigured without electrical intervention. The general concept of software is very broad and includes operating systems (OSs), utilities, drivers, firmware, middleware, application program interfaces (APIs), applications, database management systems (DBMS) and networking tools that manage data, interface between systems, provide user interfaces to hardware and process transactions on behalf of the user.

Areas of concern associated with software include:

- Logic flaws or semantic errors ("bugs")
- Lack of patching
- Inadequate or ungoverned access control
- Disclosure of sensitive information
- Improper modification of information
- Loss of source code
- Lack of version control
- Processing of input and output without validation
- Unnecessary services and protocols running on the system
- Unnecessary ports open on the system
- Lack of a security software development life cycle

Operating Systems

The OS is the core software that allows the user to interface with hardware and manages all system operations. The OS manages access to system resources and controls the behavior of system components.

Areas of concern associated with the OS include:

- Unpatched vulnerabilities
- Poorly written code (buffer overflows, etc.)
- Complexity
- Misconfiguration
- Weak access controls
- Lack of interoperability
- Uncontrolled changes
- Unnecessary services and protocols running on the system
- Unnecessary ports open on the system

Currently, most desktop computers use versions of Windows, Mac OS or Linux, while mobile devices commonly run Apple iOS or Google Android. These systems are subject to periodic version updates that may affect application programming interfaces (APIs) and subsystems in ways that can impact third-party applications. This applies to mobile phones as well as desktop or laptop computers, where access to APIs may be available to third-party developers with minimal oversight. However, the greatest risk arises from unsecured access to APIs within an organization's own network, creating the potential for code to be written or moved into production without proper oversight.

Upgrades to organizational devices under central management may be managed to avoid real-time updates until proper testing has been done. However, personal devices may update automatically, or users may choose to update them as soon as possible to gain access to new features. As with hardware, risk practitioners should keep in mind that any reliance on the OS capabilities of personally-owned devices introduces the possibility of risk arising from changes or impacts outside organizational control.

The risk practitioner should keep in mind that no commercially available OS is absolutely secure; therefore, regardless of the underlying platform, it is necessary to implement strict controls over changes, patches and configurations. Where practical, systems should be hardened to disable unnecessary services and default accounts and passwords should always be changed.

Deployment of security patches is important and should be subject to a patch management process that identifies, tests and schedules implementation. Testing should not be skipped in an effort to expedite patching, because it is possible for a patch to create negative effects on systems. Even with rigorous testing, after a patch has been implemented, it should be monitored to ensure that all related systems, utilities and applications are still working correctly.

Applications

Applications are the face of the information system and are the mechanism by which most users can access information, perform transactions and use system features. Most functions are accomplished by applications, which interact with the OS using APIs or other programming methods. When applications are negatively impacted, the result is often a loss of value to the enterprise.

Areas of concern associated with applications include:

- Poor or no data validation
- Exposure of sensitive data (i.e., lack of encryption/obfuscation)
- Improper modification of data
- Logic flaws (i.e., logic errors)
- Software bugs (i.e., coding errors)
- Lack of logs
- Lack of version control
- Loss of source code
- Weak or lack of access control
- Lack of operability with other software
- Back doors
- Poor coding practices
- Lack of governance over how data are stored

Application risk is often due to flaws or bugs in the coding of the application. This is especially true for web and mobile applications, including those developed by an organization for internal use as well as those written by third-party developers. Countermeasures to web application vulnerabilities are related to proper design, coding and testing. The Open Web Application Security Project (www.owasp.org) and similar resources list common web application vulnerabilities and provide direction on how to mitigate and test for such vulnerabilities. See chapter 2 IT Risk Assessment for more information.

Databases

Where large amounts of information need to be stored and retrieved on a regular basis, organizations commonly use databases. Database security is important because databases tend to contain critical and sensitive information that supports business operations.

Database security is provided through the following:

- Encryption of sensitive data in the database
- Use of database views to restrict information available to a user
- Secure protocols to communicate with the database
- Content-based access controls that restrict access to sensitive records

- Restricting administrator-level access
- Efficient indexing to enhance data retrieval
- Backups of databases (shadowing, mirroring)
- Backups of transaction journals (remote journaling)
- Encryption of backups
- Referential integrity
- Entity integrity
- Validation of input
- Defined data fields (schema)

Database servers should be placed behind significant protective measures and not directly accessible from outside organizational boundaries.

Software Utilities

Various forms of software utilities, including device drivers and application middleware, provide the vital bridges between hardware and applications. Drivers translate application commands into formats that hardware can understand, and like all translation, this is a trusted role. Compromised drivers may interfere with proper operations of hardware by altering intended commands to yield different results than anticipated. Corrupted drivers may also fail to deliver any instructions at all.

Areas of concern associated with utilities and drivers include:

- Use of outdated drivers
- Unavailability of drivers
- Unpatched drivers
- Use of insecure components (older encryption algorithms)
- Unpatched vulnerabilities

4.2.3 Environmental Controls

Environmental controls include power and heating, ventilation and air conditioning (HVAC) systems that support IT operations. Unlike humans, computers do not provide for their own operational subsistence; when deprived of environmental services, computer systems may immediately shut off (as with loss of power) or endure under conditions that can lead to electrical or mechanical errors (as with loss of ventilation or unregulated temperature). These impacts may in turn affect business processes by interrupting processes or causing them to be completed incorrectly.

Areas of concern associated with environmental controls include:

- Power interruptions
  - Loss of power
    - Surge
    - Spikes
    - Sags
    - Brownouts
    - Faults

- Generators
  - Insufficient capacity
  - Poor maintenance
  - Latency of generator to power on
- Batteries
  - Poorly maintained
  - Outdated
- HVAC
  - Overheating
  - Humidity problems
    - Corrosion and condensation (high humidity)
    - Static (low humidity)
  - Clogged filters
  - Lack of maintenance
- Water
  - Loss of water needed for cooling systems
- Health and safety issues
- Secure operational areas
  - Restricted access to server rooms and wiring closets
  - Secure access to power supplies, generators, elevator shafts

4.2.4 Networks

Much of the value associated with computing power today arises from bringing information from one place to another for collaboration, distribution, efficient processing or other purposes. When two or more computers are brought into connection with one another, the result is a network.

Networks take many forms and can be made up of a variety of specialized devices including cabling, repeaters, switches, routers, firewalls, gateways and wireless access points. They can range in size from two devices to the Internet itself, which is simply a global network of interconnected devices operating on essentially the same principles. Although some networks still use technology based on continuous waves (commonly called analog networks), most modern networks are digital, where the communications are sent via electrical signals or pulses of light.

Digital signals tend to be much more resilient to errors than analog technologies, because electrical or light transmissions operate on a binary code basis (0 or 1) rather than having to account for what may be significant wave variation. They do suffer more strongly from attenuation (loss over distance) than analog traffic, because anything that becomes too weak to be detected is assumed to be a binary 0. However, this simplicity provides a correlating benefit: whereas analog attenuation tends to introduce noise that makes rejuvenation difficult, any detectable signal on a digital line can be immediately boosted to full strength with confidence that it represents a binary 1. As a result, even the faintest digital signal can be rejuvenated without introducing errors, provided that this rejuvenation occurs before the signal becomes too weak to detect. These traits make digital networking both simpler and faster than analog networking.

CHAPTER 4— INFORMATION TECHNOLOGY AND SECURITY

Networks warrant special consideration in IT risk management. Partly, this is because they are often the targets or channels used to attack systems or applications; however, the network itself is often essential to business operations, and many business processes rely on its continued availability as a precondition for most business processes.

Networks are used for many purposes, including:

- Transferring data between individuals
- Transferring data between applications
- Controlling and monitoring of remote equipment (such as industrial or supervisory control and data acquisition [SCADA] networks)
- Backing up data
- Enabling communication between end-user devices (e.g., Bluetooth)

When assessing networks, the risk practitioner should consider:

- Network configuration and management, including the criticality of network operations
- Network equipment protection
- The use of layered defense (defense in depth)
- Suitable levels of redundancy
- Availability of bandwidth sufficient for intended use cases
- Use of encryption for transmission of sensitive data
- Management of encryption keys
- Damage to cabling and network equipment
- Eavesdropping on communications through physical tapping or logical interception (e.g., man in the middle)
- Choices and documentation of network architecture
- Management of flow of information via network devices like firewalls, routers, switches, etc.

Protocols

Networks send signals using a variety of physical methods. However, all of them have the same result: data sent by one participating device is delivered to another to process. These data typically comprise messages but are broken into small pieces (packets) to allow multiple messages to traverse networks at the same time. For applications to receive data successfully, a series of intermediary steps is needed to interpret the binary code received by the destination host. Specifically, what portion of the received data are data that were originally sent, versus data that were added on to help the transmission reach its destination?

Destination hosts can answer this question by knowing in advance what format the received data are meant to take, which is a function of how it was packaged and processed. These rules for packaging and processing of network traffic are called protocols.

The TCP/IP Stack

Numerous specialized protocols exist; however, one set of protocols has gained prominence to the point of supremacy. That is the protocol suite commonly known as Transmission Control Protocol/Internet Protocol (TCP/IP)—just two of more than a dozen distinct packaging and processing forms that exist within the suite.

What makes TCP/IP so popular and special is that it was designed to implement the sorts of capabilities needed to make networking work on a massive scale. Indeed, as the name implies, the IP is the backbone of the Internet.

TCP/IP is based on a reference protocol set called the Open Systems Interconnection (OSI) model. It does not implement the OSI model precisely as originally envisioned; the OSI model has seven layers, while TCP/IP has five. However, TCP/IP does provide for all of the functions specified in the OSI model.

At the base of TCP/IP is **Layer 1, the physical layer**. Here, physical infrastructure is put into place to carry transmissions, which may be analog or digital in transit but will ultimately be translated into digital signals at the endpoint. The most common local-area protocol for physical transmission is Ethernet, defined in IEEE 802.3; for long-haul systems leveraging optical fiber, Fiber Distributed Data Interface (FDDI) is often the protocol used. These protocols define the network hardware.

Above the physical layer is **Layer 2, the data link layer**, which handles the transfer of data across network media—how signals are sent and received. Traffic processed at Layer 2 is addressed for specific recipients using addresses coded into the hardware, called media access control (MAC) addresses; however, Layer 2 networks do not know where to find specific MAC addresses. Instead, Layer 2 relies on a broadcast model in which messages are delivered to all local nodes and then discarded by those endpoints that do not match the MAC address. The Point-to-Point Protocol (PPP) is a Layer 2 protocol.

Next comes **Layer 3, the network layer**. The purpose of the network layer is to move messages from one network to another, and its principal protocol is IP, which uses addresses that specify destinations in ways that indicate relationships between participant networks. IP works closely with **Layer 4, the transport layer**, where data transfer and error checking capabilities are implemented. Messages that require high fidelity are generally sent using the TCP, which is connection-oriented and seeks retransmission of lost or damaged message segments. Messages whose processing is time-sensitive may instead be transferred using the User Datagram Protocol (UDP), which operates on a "best effort" basis and allows for some degree of loss in transit.

The top layer of the TCP/IP stack is called the **application layer**, commonly known as **Layer 7** because it combines the three top layers of the seven-layer OSI model into one. (The other OSI layers are the session and presentation layers.) Network protocols that operate at Layer 7 include the Domain Name System (DNS), Network File System (NFS), Lightweight Directory Access Protocol (LDAP) and Simple Network Management Protocol (SNMP). Messages originate as application data on the sending host and pass down the stack to Layer 1, with header data added according to the particular protocols at each layer as shown in **figure 4.2**.

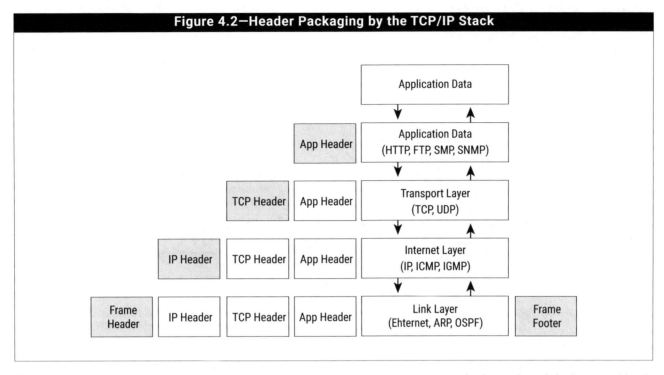

Figure 4.2—Header Packaging by the TCP/IP Stack

The physical layer allows the data to flow as electricity, light or sound between endpoints, where it is processed back up the stack to reach Layer 7 on the destination host. Here, received messages are acted upon by the destination host's high-level software.

Cabling

Even as wireless networking has proliferated, physical links between systems continue to be of primary importance for many applications, including connecting devices subject to considerable interference and establishing high-speed links between core devices. Cabling comes in several forms including unshielded twisted pair (UTP), coaxial, and optical fiber. Each type of cable has its own benefits and vulnerabilities, and the risk practitioner should ensure that the appropriate cabling is being used for the communications required.

UTP is graded is according to its specific speed ratings and attenuation potential. The least expensive option used for many local area networks (LANs) is UTP cable with a grade of Category 5e (CAT5e) or Category 6 (CAT6). Category (CAT7), a shielded cable that protects each pair of wires and the cable itself to reducing noise and cross talk for ultra-high speed is available, but not yet in common use. For connections between networks (such as Internet connectivity), cabling is typically either coaxial or optical fiber, depending on the provider.

Areas of concern for the risk practitioner regarding cabling include:

- Physical security of cabling
- Cable exceeding the approved length of the cable runs (100 meters for CAT5e, 55 meters for CAT6)
- Protection from damage to cabling (conduit)
- Use in an area of high radio frequency interference (RFI), which may require shielding
- Use of cable that is not of suitable standard (e.g., CAT3)
- Ensuring use of plenum-rated cable where required
- Improper terminations of cable on connectors
- Lack of cabling records

Cabling is part of the infrastructure that defines the physical layer (Layer 1).

Repeaters

Repeaters are used to extend the length of a signal being transmitted over cable or wireless networks. Regarding the advantages of digital over analog networking, signals suffer from attenuation as they traverse distance. Repeaters placed at properly offset locations are able to receive an attenuated input signal, regenerate the signal to full strength, and send it on further. The distance between repeaters is based on the type of cabling or technology being used and the operational environment. For example, as the quality of fiber has increased over the years, it has meant that the distance between repeaters has grown substantially. In an environment with many concrete walls or other interference, or where a higher frequency wireless channel is being used, repeaters may be required more often.

One risk associated with repeaters is that separating them by too much distance may fail to provide a clean, error-free signal. Another is that a wireless repeater providing a strong signal into areas outside the perimeter of the organization's facilities might be subjected to unauthorized access.

Repeaters operate at the physical layer (Layer 1).

Switches

Switches are omnipresent in networking. In all forms, they are used to connect devices together, and there are switches that operate at Layers 3, 4 and 7 to handle functions such as load balancing. However, their most common function is to provide dedicated pathways to endpoints based on MAC address association (a Layer 2 function).

In addition to connecting devices, switches can also segment and divide networks by isolating the ability of traffic to flow between endpoints. With such configurations in place, the results are comparable to running distinct physical infrastructure.

Areas of concern associated with switches include:

- Physical protection
- Ensuring proper configuration
- Documentation
- Device failure
- Protocol attacks

Routers

The purpose of a router is to connect multiple networks together and forward incoming packets in the direction of the destination IP address that is in the packet header. The router knows where it might forward packets based on a table of values that it builds using routing protocols, such as the Border Gateway Protocol (BGP); by updating these tables based on messages that it receives, the router learns how best to direct traffic toward its destination.

In many ways, a router is like a traffic circle (roundabout) used to handle traffic on roadways. Traffic enters the circle from any point and leaves following the road that is best aligned with the vehicle's ultimate destination. The road taken may not be a direct pathway, but it will move traffic towards the destination. Routers regularly receive updates from their peers to provide awareness of network conditions, and a router may direct traffic to different routes depending on traffic volumes and congestion.

Some network traffic has labels that indicate it has a priority, similar to the way that emergency vehicles are permitted to bypass other traffic. These labels indicate that the packets should be processed quickly and not be held back by other packets waiting to be processed. The delay in processing is called latency, which might severely affect the quality of some types of processing, such as voice-over IP (VoIP) telephone calls or video teleconferencing. Labeling of the traffic allows for provision of traffic management based on target quality of service (QoS) and class of service (CoS). QoS is usually used to provide a guaranteed bandwidth for traffic volumes, while CoS is used to grant priority to certain packets over others (e.g., voice packets over regular data packets). This also helps to prevent jitter, a variation in the arrival time of a packet.

In addition to sending traffic on its way, routers can also be configured to block certain kinds of traffic based on source address, protocol or other characteristics present in the IP header. Blocking traffic is a secondary role for routers, which are intended primarily to ensure that traffic moves towards its destination. However, limited blocking may be part of a layered defense strategy to protect networks.

Areas of concern associated with a router include:

- Improper configuration
- Use of weak protocols (e.g., RIPv1)
- Software bugs
- Unpatched systems
- Physical security
- Unintentional support for IPv6 (e.g., Teredo tunneling capabilities turned on by default)

Routers operate at the network layer (Layer 3).

Firewalls

The term "firewall" refers to a variety of technologies that operate at various layers of the TCP/IP protocol stack and have evolved significantly over the past years. However, in all forms, a firewall is a system or combination of systems that exists to enforce a boundary between two or more networks, typically between a secure environment and an open environment such as the Internet. Firewalls may be used to control both incoming and outgoing traffic, and this is desirable in many cases. Basic types of firewalls are described in **figure 4.3**.

	Figure 4.3—Firewall Types
First Generation	A packet-filtering router that examines individual packets and enforces rules based on addresses, protocols and ports, without awareness of how any given packet relates to any others.
Second Generation	Operates at Layer 4 and keeps track of connections in a state table, which allows it to enforce rules based on packets in the context of the communication session; commonly known as a stateful-inspection firewall.
Third Generation	Operates at Layer 7 to examine the application-level protocol being used, such as Hypertext Transfer Protocol (HTTP). Sensitive to suspicious activity related to the content of the message rather than being limited to the address information.
Next Generation	An enhancement to third generation firewalls that brings in the functionality to prevent intrusions by inspecting Secure Sockets Layer (SSL) or Secure Shell (SSH) connections through deep packet inspection.

Firewall configurations should backed up regularly, reviewed to ensure that all rules are in the correct order and documented. Firewalls should also be tested on a scheduled basis, and any changes to the firewall configuration or rules should be subject to organizational change management processes. Additionally, the risk practitioner should verify that the staff members managing firewalls and other network devices are knowledgeable, trained and

CHAPTER 4— INFORMATION TECHNOLOGY AND SECURITY

supervised, with firewall logs regularly reviewed to detect any suspicious activity. In some enterprises, review of firewall logs is outsourced to a security services company. In such cases, the risk practitioner should review and understand the terms of the contract governing this service.

Proxies

A proxy is a device that acts as an intermediary between two communicating parties by presenting itself to each side of the communication as if it were the host on the other end. This arrangement allows the proxy to filter and examine suspicious activity, protect internal resources and act if unacceptable activity is occurring. Proxies are typically established at network perimeters. In some cases, proxy capabilities may be built into next-generation firewalls; however, many enterprises use dedicated servers or appliances for proxy functions.

Intrusion Systems

Firewalls and perimeter routers are secured with the intent of creating a fortified perimeter through which malicious traffic will not pass. Layered defense, sometimes called "defense-in-depth," assumes that perimeters are insufficient to protect the network and places additional security inside for added assurance. Intrusion systems come in two forms: the intrusion detection system (IDS), which identifies potentially malicious traffic, and the intrusion prevention system (IPS), which combines an IDS detection capability with selective firewall features to block traffic identified as potentially malicious.

The basic rationale for using an IDS over an IPS is that not all traffic suspected of being malicious actually is, and that it is preferable to alert someone to investigate rather than cut off what might be normal or legitimate communications. This goal means that the IDS need not process traffic as it passes through but instead can sample traffic for analysis as an observer. An IDS, therefore, has no effect on network throughput. In contrast, for an IPS to work, it must be placed in the line of traffic so that it can stop it should the need arise. This generally means that an IPS will slow down a network based on the limits of its processing power.

Both detection and prevention systems can be implemented at the network or host level. In the latter case, the IDS or IPS takes the form of a software application and can be set up to monitor activity in the general sense or specific targets of interest, such as attempts to change system files.

The Domain Name System

Network addresses exist as sets of numeric values that exist in hierarchies. Requests to access distant endpoints are processed based on these addresses. However, most people so not know the addresses associated with particular endpoints or websites. In that sense, the DNS is the mechanism that makes the Internet work. DNS provides a simple cross-reference that is used to associate a written name with the IP address used by network devices. An individual or organization registers the name for a website and provides some basic technical information, after which users who want to visit that site can type the name instead of the address—for instance, www.isaca.org instead of 104.16.215.248. Not only is it more convenient to remember the name than the address, but later changes to the address need not affect the ability to reach the website so long as the mapping remains up to date.

DNS is structured as a logical tree. When a network device does not know the IP address associated with a particular name, it sends a DNS request up through the tree to a higher-level DNS resolver. At the top level are root DNS servers that can direct queries back down appropriate paths to servers that have the answer. The reply is then sent to the requesting device using UDP over port 53, after which the requesting device stores the IP address and name to reference for future use.

There have been many attacks on the Internet using DNS over the years. Malicious users have sent false DNS replies to misroute traffic, and DNS replies have been used in amplification attacks to flood a victim's system. Attacks against DNS servers that impede name resolution can cause considerable difficulty for e-commerce and other basic business transactions. DNS can also be used to learn information about a company that may be useful in planning attacks. Risk practitioners reviewing networks should take time to understand their enterprise's specific approaches to DNS and the security features that have been put in place to limit these and other DNS-related risk.

DNS is an application protocol that operates at Layer 7.

Wireless Access Points

Wireless devices offer a level of flexibility and ease of use not possible with traditional cable-based networks, allowing users to move around and log in from multiple locations without requiring network cables or ports. However, the lack of a defined physical boundary increases the threat of unauthorized users being able to access wireless-enabled network.

Wireless infrastructure is typically established using one of more of the protocols associated with the 802.11 standard, with specific standards allowing for specific maximum transfer speeds, security protections and other features. A good practice is to segment the wireless infrastructure into a network distinct from the wired infrastructure and implement strong password requirements for network access. Placement of wireless access points is also important. Risk practitioners reviewing wireless networks should ensure that access points are in locations not subject to interference from other devices. Access points operating on WiFi 5 (802.11ac) or later are designed to moderate their power levels to avoid creating interference over distances and often deliver the best performance when located inside the rooms that they will support. Placing access points above the ceiling may result in signal degradation arising from HVAC and utility pipes.

Insertion of unauthorized (rogue) wireless access points on the network may permit a person to access the network directly without having to connect through a secure device or to bypass firewalls or other layers of defense. Detecting this type of vulnerability is generally possible only with a combination of diligence and technical expertise, particularly if the rogue access point has been installed by someone who has legitimate physical access to the organization's workspace. The risk practitioner should ensure that wireless discovery scans are part of the overall vulnerability assessment process, especially) if wireless technology is not known to be used within the enterprise.

In addition to 802.11 network signaling, organizations may also have Bluetooth devices present in workspaces. Bluetooth was once standardized as IEEE 802.15.1 but is currently maintained by the Bluetooth Special Interest Group. Its purpose is to easily link devices physically close to one another into personal areas networks (PANs). Some of the most common uses for Bluetooth are linking computers or mobile phones to peripherals such as keyboards, mice, and headsets. Several attacks against Bluetooth have been identified by researchers as viable; however, successful exploitation of these attacks is rare.

Network Architecture

Networks can be built in a variety of ways, known as architectures. The architecture of a network has implications for how easily data may be shared between specific endpoints and even what connections between endpoints are possible. The choice of network architecture is based on the number of devices, distance and communications requirements such as speed, confidentiality and reliability.

The architecture of an information system plays a role in determining the appropriateness of controls and other forms of risk response. For example, controls that focus on traffic at the network layer are likely to be effective in addressing threats directed against network-layer targets, but they are unlikely to be effective against application-layer attacks. Multiple controls are typically needed within an IT architecture to provide robust assurance against malicious activity.

A key factor in the maturation of the processes and practices of an enterprise is the development of enterprise approaches to risk management, architecture and business continuity. Having an enterprise approach does not mean that different divisions or business areas are forced to accept a cookie-cutter model. These areas should tailor their approach through how requirements are met, not by varying what is required. Addressing these functions at the enterprise level promotes consistency, repeatability, compliance and accountability. It also improves the visibility afforded to senior management regarding the practices and strategy of the enterprise.

Relatively few enterprises have a mature IT architecture. In many enterprises, business processes rely on systems built as part of individual projects or initiatives, and each system is an independent entity with little in common with other systems. The lack of an EA results in ownership gaps between systems and unclear areas of responsibility for incident or configuration management.

As the complexity of an architecture increases, it invariably becomes more challenging for the enterprise that owns it to secure it and ensure compliance with security standards, regulations and good practices. The risk practitioner may find that systems and networks in the organization are not mapped out on a network diagram, and some systems and remote-access capabilities tied to legacy applications may not be documented at all. Controls may overlap or conflict with one another, and there may be methods to bypass controls or single points of failure. Under such circumstances, the ability of the enterprise to measure risk or identify vulnerabilities is severely impacted. The risk practitioner should consider adoption of a coherent, structured enterprise architecture a high priority.

Network Topologies

The arrangement of endpoints on a network is known as a network topology. **Figure 4.4** illustrates the commonly used bus, ring and star topologies.

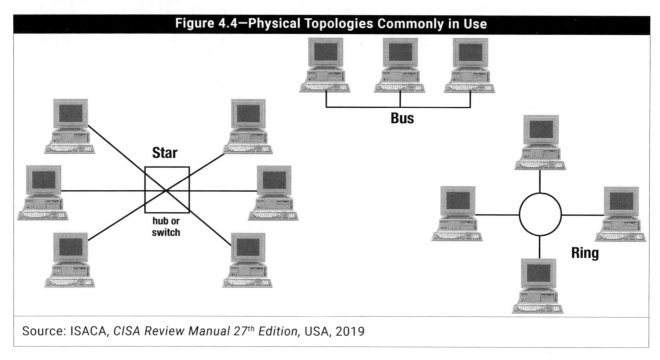

Source: ISACA, *CISA Review Manual 27th Edition*, USA, 2019

A bus topology connects every device by means of one communications path. The design is simple and cost effective, which is why it has been used for decades to carry cable television signals. The key vulnerability lies in upstream dependency: a cut cable results in failure for every user connected downstream of the cut. It is also

CHAPTER 4— INFORMATION TECHNOLOGY AND SECURITY

relatively easy to intercept traffic on a bus network, a point of particular concern for some users of cable-based Internet access.

In a star topology, every device is connected to a central switch, which is more efficient than a bus since there are shorter paths from any end point to the switch. The switch design also makes it more difficult for one user to intercept traffic intended for another. Star topologies are more vulnerable to interruption than bus topologies, because loss of the central switch affects all users. However, the switch can generally be more easily protected than a bus cable.

A series of star networks arranged with branches to other star networks is sometimes called a tree topology. Tree networks are popular because they scale well: a cut link between the branches of the tree causes isolation of that branch but does not otherwise impede the network.

A ring topology is used in backbones and areas where reliable high-speed communications and fault tolerance is desired. A ring connects every device and allows traffic to pass in one or both directions; if the ring is bidirectional, a single cut does not affect the network at all. Ring topologies are excellent for busy networks where switching would introduce too much delay due to processing overhead. They are also considerably more expensive that bus, star or tree topologies.

Where the highest availability is required, a mesh topology is the answer. In a mesh, many devices are connected to many other devices so that traffic can route around a failure in any part of the network. The Internet itself is a partial mesh and was built to survive massive failures of any one part of the network while still allowing communications over the rest. Mesh networks are very costly, and they do not scale well with physical lines due to the need to connect each device to many other devices. Wireless links make locally oriented mesh architectures considerably easier to architect and maintain.

Network Types

Certain elements of network architecture depend on how a network is intended to be used. There are two principal types of networks: local and wide area.

Local Area Networks

A LAN is a network that serves several users within a specified nearby geographic area, such as a building, floor, or neighborhood. LANs often use star or tree topologies, although wireless LANs may be meshed.

Although LANs vary with regards to both user count and services provided, the common trait of every LAN is that its endpoints are logically close together.[6] Within that basic definition, numerous types of networks can arise. LANs intended purely for personal use within homes or around individuals are sometimes called PANs, while a LAN built for the express purpose of connecting storage devices to servers and other computing devices might be called a storage area network (SAN). Despite the variations in these use cases, both are LANs for practical purposes, because traffic within these environments is within the same basic routing address block.

Wide Area Networks

When endpoints are logically distant from one another, operating on separate routing blocks, the solution is to interconnect them using a wide area network (WAN). The remote locations connected by WANs may range from physically short distances, such as floors or buildings, to extremely long transmissions such as regions or even

---

[6] This may not mean *physical* proximity; the nodes on a LAN are close to one another in terms of how they relate to one another from a routing and switching perspective.

continents. Some sources refer to WANs that extend only over limited areas such as cities as metropolitan area networks (MANs). However, their practical traits remain the same: they carry traffic destined to cross network borders.

Software-defined Networking

Networking at both the local and wide area is undergoing a revolution driven by software definition. By decoupling network control and forwarding functions, software-defined networking (SDN) allows dynamic shaping of traffic in software. SD networks can be changed dynamically to meet business requirements on short notice. For instance, wireless access points might be modified to switch between 2.4GHz and 5GHz based on network conditions, while an SD-WAN can manage connections across broadband, mobile and traditional circuits simultaneously. Virtualization of customer-premise equipment reduces cost associated with establishing or closing workspaces, allowing greater business agility than has traditionally been the case.

Demilitarized Zones

Network borders exist to prevent outsiders from directly accessing systems within particular LANs. Some services are intended for outside use, and these are typically placed in network segment off of the border, expressly configured to allow open access. All devices in this demilitarized zone (DMZ) are hardened with unnecessary functionality disabled. IDSs or IPSs are generally placed to monitor, record and potentially block suspicious activity. Firewalls in the DMZ are generally placed behind packet-filtering routers to clear out obvious bad traffic in advance.

As can be seen in **figure 4.5**, the firewall ensures that traffic from the outside is routed into the DMZ where the web server is located. Critical organizational assets are not placed within the DMZ because it is subject to attack and compromise from the outside. External users cannot access the internal network directly, thus providing the internal network with some measure of protection.

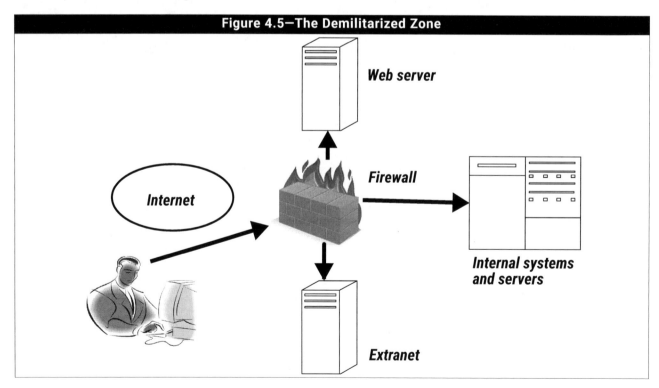

Figure 4.5—The Demilitarized Zone

Virtual Private Networks

Under ordinary conditions, it is fairly easy to capture WAN traffic passing over the Internet or by way of directional connections. Virtual private networks (VPNs) were designed to make this much more difficult. A VPN uses the existing telecommunications infrastructure without regard to its inherent security. However, the data signal is logically separated from other traffic by digitally-constructed tunnels, which are then configured using encryption (see section 4.8) to make the traffic unreadable except by intended recipients. Once established, a VPN functions in a manner similar to a point-to-point leased line but at a considerably lower cost.

Current computing architectures may support many VPN connections simultaneously, allowing a substantial portion of the organization's workforce to connect to corporate infrastructure from remote locations—potentially anywhere in the world where the Internet can be accessed, whether by wired or wireless infrastructure. Once connected, VPN participants are logically close to one another and operate as part of the same LAN. VPNs may be established using several different technologies, including the Internet layer using the security extensions to the Internet Protocol (IPSec) or the Transport layer using Transport Layer Security (TLS).[7] In general, an IPSec VPN is more commonly used for bulk channel encryption between networks, while a TLS VPN is more likely to exist between network endpoints.

4.2.5 Technology Refresh

The age and condition of technology used by an organization can present a substantial risk factor. Many organizations use outdated technologies that are difficult to obtain, support and maintain. The technology in use may have been acquired through multiple projects and/or mergers and consist of products from a varied mix of vendors, languages, configurations and vintages, and it may run on different operating systems, hardware, network architectures and databases. A large organization might have systems based on mainframes, client-servers, virtual environments and the cloud all at once. The sheer complexity that this represents is a source of risk.

In general, enterprises that retain legacy systems do so because there is a compelling business need. It is good practice to have a simple, easily managed and controlled environment with standard products and technologies, but this may be nearly impossible to attain in a large organization. Many systems may continue to be used long after their anticipated life span simply because it is not cost-effective to replace them, especially in organizations carrying out manufacturing or industrial activities.

Considerations that affect risk assessment related to technology include:

- Age of equipment
- Expertise available for maintenance
- Variety of vendors/suppliers (are they still in business?)
- Documentation of systems
- Availability of replacement parts
- Ability to test systems or equipment
- Operating environment and user expertise
- Ability to patch/mitigate vulnerabilities

The risk practitioner should work with business process owners and technical staff to determine what legacy systems are in use. It is also important verify both that their continued use represents a compelling business need and that the risk associated with such use has been documented and accepted by senior management. In many cases, legacy systems persist out of need, but managers are unaware of the extent to which the organization incurs risk from their

---

[7] TLS replaced the earlier Secure Socket Layer (SSL) protocol, and TLS VPNs are sometimes called SSL VPNs.

persistence. Under such circumstances, the role of the risk practitioner is to advise the risk owner of ways in which the assessed risk might be addressed in order to bring it into alignment with the organizational risk appetite (or at least to tolerable levels).

4.2.6 IT Operations and Management Evaluation

Staff members responsible for IT operations are frequently the first to be aware of problems relating to systems and applications. Interviewing operations staff and reviewing logs and trouble tickets can help the risk practitioner gain insight into an unmitigated or recurring problem within a system, which should be investigated to determine whether they pose unaccepted risk to the enterprise.

IT operations typically include multiple functions related to risk management and quality assurance, including policies and procedures for the management of incidents, configurations, problems and releases of new systems or software. The risk practitioner should work with IT staff to understand these functions and how they are carried out. Weaknesses in any of the IT functions may create opportunities for threat agents to exploit systems or processes.

Many elements of IT risk are a result of how an IT department is run and managed. Where an attitude of complacency reigns, risk ignorance can result in breaches of security. Security problems commonly seen within the IT department include sharing high-level accounts and passwords and the lack of control over changes to production systems. The pressure to complete projects on time or meet demands from the business can result in shortcuts that pose significant unaccepted risk to the business.

Configuration Management

Devices and systems must typically be configured after installation to support communications, interfaces with other organizational systems and secure operations directed by policy. The goal of configuration management is to have a single approved way in which systems or devices intended for a particular role should be configured for use. Use of a common configuration makes it easier to deploy new systems, test patches and upgrades, identify the presence of malware, and manage the enterprise.

Configuration management often includes hardening, a process that reduces vulnerability by limiting the attack vectors that might be used as points of compromise. A hardened system is one that does not store any sensitive data that are not immediately needed to support a business operation. In addition, it has all unnecessary functionality disabled, including ports, services and protocols that are not required for the intended use of the system within the enterprise environment. Many devices and systems come with guest accounts or default passwords that should be changed or disabled as part of the hardening process. While hardening is a good practice, the risk practitioner should understand that it can be difficult to identify necessary services and functionality. Hardening systems based on third-party guidance should be done with great care to prevent impacts to operations. Many custom-integrated systems cannot be hardened without significant negative effects.

Configurations and settings that control the operation of devices should be backed up and available in case of equipment failure. The risk practitioner should ensure that there are policies and procedures in place that apply to proper configuration management and that backups of standard configurations are available for use in recovery scenarios.

4.2.7 Virtualization and Cloud Computing

Beginning in the mid-2000s, massive expansions in computing power combined with advances in software made it practical to create virtual machines (VMs)—instances of emulated hardware that existed in computer memory and could do everything physical computers could do, including run operating systems and applications. Establishing

CHAPTER 4— INFORMATION TECHNOLOGY AND SECURITY

servers in particular as VMs offered tremendous advantages for continuity and recovery, because these environments could be transferred to different physical systems without needing to interrupt operations. In many cases, multiple VMs could be run on single physical servers built and scaled for this purpose, resulting in significant downsizing of physical space requirements for organizational data centers.

By 2010, major technology companies were taking this a step further, offering hosted environments where virtual machines could be provisioned against readily available hardware on demand. Establishing massive data centers equipped with state-of-the-art redundancy for power, connectivity and other risk factors, these companies made it possible for enterprises that had already reduced their physical server footprint to shift what remained of their physical infrastructure to provisioned capacity accessible via remote desktop interfaces. Without having to worry about the physical element of where or how these systems were supported, enterprises could realize substantial cost savings while gaining access to capacity and protective measures that represented the best in the industry. The opaque nature of what occurred within these Internet-connected data centers led to the practice to be commonly called "cloud computing," referring to the use of a cloud in network diagrams to refer to the Internet.

Consolidation and virtualization reached truly transformative levels with the advent of hyper-converged infrastructure (HCI), a software-defined system for unified data center operations. HCI combines storage, processing, networking, and management functions that run on x86 platforms. This transformation shifts administration attention from systems to location-agnostic workloads, similar to the way that public utilities operate. Through simplified management and reduced complexity, organizations can better focus on delivering information services rather than technologies.

Virtualization has also experienced significant advances with the rise of containers. Whereas a VM is a complete replica of a physical machine, its operating system and its application software, a container is a way to establish isolated instances of application software drawing on the same operating system and potentially sharing data libraries. Containers can be built and deployed much faster than VMs and consume fewer resources on an individual basis; they can also be built *from* virtual systems. By avoiding the need to replicate the full hardware and operating system functionality in a distinct instance for every instance of the application, containers drive down cost and improve performance while retaining virtual isolation in most regards. The ability to package containers promotes application portability across underlying platforms and promotes software development based on Agile methodologies.

Cloud computing has many advantages; however, it also creates risk. One area of concern is that, while cloud data centers are exceptionally protected, they are also exceptionally tempting targets because of the sheer volume of data that consolidation brings together. The specific security practices of a given cloud provider may not be known to an enterprise contracting with them when the provider shares only a standardized report from a third-party auditor assessing compliance with certain standards. Enterprises should also take care to understand where their data may be physically stored, since this can determine jurisdiction for purposes of privacy law and regulation (see section 4.10 Data Privacy and Data Protection Principles).

Risk practitioners whose organizations are considering a migration of infrastructure to the cloud should keep in mind that this is a business decision. However, outsourcing of any sort is a form of risk transfer, and as discussed in chapter 3 Risk Response and Reporting, enterprise cannot fully transfer their risk. Should organizational data be compromised, or systems fail in cloud environments, the enterprise will experience the bulk of the impact, regardless of what offsets may be afforded by service level agreements or contractual terms. These considerations may not alter the decision to migrate to the cloud. The risk practitioner should ensure that they are included in the discussion.

See section 2.3.1 Sources of Vulnerabilities for more information on cloud computing vulnerabilites.

4.3 Project Management

A series of tasks that work towards a common purpose with a defined end-state is called a project. Projects are undertaken in every enterprise and field of endeavor. The goal of a project is to deliver value by bringing about some specific set of conditions or outcomes.

Enterprises rely on projects and programs to execute the decisions of management and carry out their business functions. Projects may also be grouped with long-term programs grouped into portfolios that share common characteristics. The goal of portfolio management is to optimize business value and create a greater degree of oversight under one accountable authority who is also appropriately empowered to solve problems as they arise.

Effective project management relies heavily on proper definition of a project at its inception and during the earliest stages of planning. Clearly defined objectives and requirements help to contain the scope of a project within boundaries that are aligned with agreed-upon resources. Clarity also greatly increases the likelihood that what a project delivers is closely aligned with what was originally requested. However, risk practitioners should keep in mind that it is not uncommon for requested features to prove less useful or important in practice than anticipated at the onset of planning.

Methodologies that emphasize regular customer evaluation of partially-complete deliverables are more likely to identify these variances between real and anticipated need prior to finalization of deliverables, which can help manage risk. Agile, Kanban and Extreme Programming (XP) are all examples of methodologies that incorporate this style of interim evaluation, while the waterfall methodology is most commonly associated with delivery of precisely what was originally requested. Which methodology is most appropriate for a given use case is likely to vary based on the nature of the project, customer expectations, organizational culture and other factors.

4.3.1 Project Risk

There is considerable risk associated with the management of projects and programs. Because projects have clear end-states, they also have the clear capacity to succeed or fail. Project failure is surprisingly common and may take the form of early cancellation, non-delivery of expected outcomes or lack of expected value relative to invested resources.

IT projects fail for various reasons, including:

- Unclear or changing requirements
- Addition of requirements without adequate resourcing (scope creep)
- Inadequate access to resources, including funding and specialized skill sets
- Problems with technology
- Delays in delivery of supporting elements/equipment
- Unrealistic timelines (push to market)
- Lack of progress reporting or effective communication

Project management is the formal discipline of organizing, administering and carrying out projects. There are several distinct methodologies of project management, not all of which have a specific individual with the title or role of project manager (PM). However, even where there is no individual PM (as in self-directed methodologies such as Agile), the behavior of project teams is always intended to bring about the same effects as would arise from having a dedicated PM, with the intent of bringing projects to successful completion.

CHAPTER 4— INFORMATION TECHNOLOGY AND SECURITY

Project management is, therefore, partially a form of risk management focused on projects. When a project is at risk, it is important to identify the root cause of the problem and take steps to address it as soon as possible. Lack of good project management can lead to:

- Loss of business or competitive advantage
- Low morale among staff members
- Inefficient processes
- Lack of testing of new systems or changes to existing systems
- Impact on other business operations
- Failure to meet contractual requirements
- Violations of law or regulations
- Insufficient QA activities

Projects are subject to numerous types of risk. Fortunately, project management methodology is well-developed, and many of the common types of risk also have common responses, such as:

- Implementing a change control board to prevent scope creep
- Prioritizing critical project tasks to use resources optimally
- Reorganizing or providing additional resources to overcome bottlenecks
- Replacing or supplementing project managers who fall short of expectations
- Canceling projects whose costs or schedules are far outside of projections or restarting them under more favorable conditions
- Replacing suppliers or renegotiating contracts that fail to deliver agreed-upon products or services

Timely identification of a project at risk of failure is important so that corrective action can be implemented while the situation can be salvaged. Identifying project risk is challenging in part because it can be difficult to obtain accurate data on project status and identify root causes of observed problems. Proper and engaged oversight and monitoring of the project along with validation of progress reports may reduce or even prevent inaccurate project status reporting. Objective risk assessments are often instrumental in determining true project status. It is a fundamental law of project management that time, quality and cost exist in a balance. Calls to "do more with less" are workable only in environments where extreme levels of inefficiency or waste are both present and correctable, which is uncommon. Far more often, quality absorbs the impact. For instance, a project team that is struggling to remain on time and within budget while maintaining access to necessary resources may skip critical steps (such as testing) as a means of attempting to balance these priorities. Such actions create a temporary illusion of success but may lead to production failures, inaccurate reporting and data processing or missing functionality as the project is implemented. The organizational impacts of cutting corners can include both direct and indirect financial losses, especially in terms of reputational impact that limits opportunities for future growth or revenue.

Projects are frequently assembled into programs. A program is a structured grouping of interdependent projects and steady-state operations that is both necessary and sufficient to achieve a desired business outcome and create some specified value. It is common for a program to rely on each project being completed successfully on schedule. Accordingly, when a project is at risk, the entire program may also be at risk.

Risk practitioners who identify deficiencies in project planning or execution should bring these to management's attention through portfolio channels as soon as possible, because project trajectories become increasingly difficult to correct as they move further away from expectations. In particular, the risk practitioner should be aware of two specific concerns associated with project and program management: first, that a project may not meet its objectives, and second, that the failure of one or more projects may affect the performance of a program.

4.3.2 Project Closeout

Projects are distinguished from programs by their finite life span. A project may last months or years, but there must be an anticipated point at which deliverables are fully transitioned to users and/or system support staff and the project is closed.

Formal closeout procedures are an important part of managing risk. In particular, post-implementation or after-action reviews allow identification and capture of lessons learned that can be used to improve future projects. Stakeholder satisfaction should be considered as part of this process. However, projects are typically undertaken for the benefit of their sponsors. Accordingly, verifying sponsor satisfaction with outcomes is particularly vital.

Closeout procedures are often standardized within large enterprises. **Figure 4.6** lists five steps that are frequently addressed.

Step	Action
	Figure 4.6—Project Closeout Steps
1	Assign responsibility for any outstanding issues to specific individuals and identify the related budget for addressing these issues (if applicable).
2	Assign custody of contracts and archive or transfer documentation to those who need it.
3	Conduct an initial post-implementation or after-action review with the project and development teams, users and other stakeholders to identify lessons learned that can be applied to future projects. Topics to cover include: • Fulfillment of deliverable targets and any additional objectives • Attainment of project-related incentives • Adherence to the schedule and costs • Analyses of team dynamics and communication with outside stakeholders • System logs
4	Document in the risk register any risk identified in the course of the project, including risk that may be associated with proper use of the deliverables.
5	Complete a second post-implementation review after sufficient time has passed for the project deliverables to realize their full benefits and costs and use the results of this review to measure the project's overall success and impact on the business.

4.4 Enterprise Resiliency

Enterprises encounter threat events on a near-constant basis. Some of these threat events present challenges to which the enterprise has no vulnerability. Others are met by controls or countermeasures, while others create consequences. Under the right circumstances, a threat actor may create effects that have real business impact, which may range from minor to catastrophic.

Effective risk management considers the full spectrum of these possibilities with the goal of creating resiliency across the enterprise. Resilient enterprises are not necessarily unaffected by threat events; instead, their nature and structure lend themselves to minimizing harmful impacts and returning to normal on an expedited basis. Two fundamental disciplines associated with enterprise resiliency are business continuity and disaster recovery.

4.4.1 Business Continuity

All systems fail. Specific durations and scopes of failure vary, but failure itself is assured. The purpose of business continuity planning is to enable a business to continue critical services in the event of a disruption, up to and potentially including an interruption on a disastrous scale. Rigorous planning and commitment of resources is necessary to adequately plan for such an event, and the first step in preparing a new business continuity plan (BCP) is to identify the business processes of strategic importance, meaning those key processes that are responsible for both the permanent growth of the business and for the fulfillment of the business goals. BCP development should be supported by a formal policy that states the enterprise's overall target for recovery and empowers those people involved in developing, testing and maintaining the plans.

The risk practitioner is typically not responsible for BCP development, but they are likely to work closely with the BCP teams. This is because BCPs are controls that reduce the risk of losing critical processes to acceptable levels. In particular, the risk practitioner may be involved in assessing the adequacy of BCPs, which depends on the levels of risk that they are meant to address. As with any other asset, the risk associated with a particular process depends on the magnitude of impact to the business should it be interrupted and the probability of an interruption. Thus, the result of the risk assessment should be the identification of the following:

- The human resources, data, infrastructure elements and other resources (including those provided by third parties) that support the key processes
- A list of potential vulnerabilities (the dangers or threats to the enterprise)
- The estimated probability of the occurrence of these threats
- The efficiency and effectiveness of existing risk mitigation controls (risk countermeasures)

BCPs may be established at the enterprise level, by each department or for each process. When there is an overarching plan, it is usually divided into multiple annexes that relate to the division and process-level activities that must be carried out to continue operations. It is highly desirable to have a single integrated plan to ensure proper coordination among various components and increase the likelihood that resources are used in the most effective way. Distribution of continuity planning into separate documents across an enterprise may lead to gaps that cause failures during real-world interruptions. However, where enterprise are divided into units that have relatively high autonomy or distinct business goals, distinct BCPs may be a reasonable approach.

The goal of a BCP is to provide a sufficient level of functionality in the business operations immediately after encountering an interruption so that the enterprise is able to continue as a viable entity. By definition, the continuity environment will afford a reduced level of functionality, because the enterprise will need to be concurrently seeking restoration of its regular production capabilities. For this reason, the BCP includes the continuity procedures determined by the enterprise to be necessary for the enterprise to survive, and limit the consequences of business interruption to levels that can be absorbed. Understanding the requirements of these processes includes considering both the technology and the people needed to carry out the continuity procedures. The recovery of critical business processes may be through an alternate process, including:

- Manual processing for something previously automated
- Outsourced support
- Use of on-hand inventory in lieu of production
- Use of alternate facilities
- Displacement of less critical functions on remaining capacity

CHAPTER 4— INFORMATION TECHNOLOGY AND SECURITY

Recovery Objectives

Business continuity planning begins with the business impact analysis (BIA), which identifies the critical services and products associated with value creation. The BIA also establishes the recovery point objective (RPO) for each process, which defines how much data can be lost in recovery, and recovery time objective (RTO), which establishes how quickly the process must be accomplished.

RTOs tend to be driven by how critical processes are to value creation, while RPOs reflect how dependent those processes are on records of prior iterations. For instance, in some financial services systems, recovery from an outage might be permitted to take as long as four hours. However, it might be unacceptable for any transactions up to the moment of the outage to be lost. In this case, operational systems would need to have a parallel transaction log that guaranteed that the incremental changes from the time of the last full backup could be reconstructed in the recovered system, and IT would be tasked with making sure that both the technical recovery and this reconstruction process would be finished by four hours from the time of the outage.

Contrast this example with an appliance-manufacturing operation based on a production line. Every day, the line produces appliances, and whenever it is stopped, value is not being created. That means that the RTO will be very low—potentially as low as a few minutes—with backup systems ready to take over almost instantly in the event of an outage. However, if the manufacturing is sustained over time, there may be no need to refer to how many appliances have been created at a given point in time. The system might record milestones of every 500 appliances completed, and between these milestones, there might be no record. In this case, the RPO would be as high as 499, with the understanding that, independent of the technical recovery, someone can jsimply count the boxed appliances waiting to be loaded.

Environments with low RTOs typically have highly- available system architectures (such as clustered servers), while those with low RPOs have highly-available data architectures or constant transaction logs. Virtualization and cloud computing have made it easier and cheaper to establish high availability; however, doing so continues to be among the most complex and costly endeavors of modern IT. Understanding which processes warrant high availability and which can accommodate longer RTOs and/or RPOs is an important part of effective business continuity and disaster recovery planning.

Risk practitioners should understand that acceptable outages and data loss are not endorsements of overall process importance but instead refer to importance in a short time. Certain processes may also be more or less important at certain times of the month or year, and planning should take this variable prioritization into account where possible. For instance, the internal audit function of a financial services company is incredibly important but not continuous. In most cases, audit processes can be delayed to later tiers of process recovery. On the other hand, in the weeks leading up to an external audit, internal auditing takes on outsized importance, as enterprises look to address potential weaknesses before they might be disclosed to investors or other stakeholders.

Because the BCP is based on the BIA, the risk practitioner may want to first determine how the BIA was developed in order to validate that it is accurate and considers all relevant risk factors. In organizations that have adopted a comprehensive business continuity management system (BCMS), the risk practitioner may benefit from having direct access to BCPs for review and reference in the context of risk management.

For more information on preparing a BIA, refer to chapter 2 IT Risk Assessment.

4.4.2 Disaster Recovery

While business continuity is concerned with maintaining key operational processes under reduced levels of unavailability, disaster recovery (DR) refers to the reestablishment of business and IT services following a disaster or incident within a predefined schedule and budget. The term is commonly associated with recovery from an IT perspective; however, disaster recovery can be considered a relative of business continuity due to the significant influence that recovery times have on sustaining continuity environments. Because continuity operations cannot continue indefinitely, an effective plan for recovery and reconstitution is essential to enterprise resiliency.

Timeframes for recovery are specified in the disaster recovery plan (DRP) based on the cost and length of outage that management is willing to accept, which in turn flows from the RTOs established in the BIA and built into the BCP. The risk practitioner should review the DRP to ensure that it is up to date and reflects current risk scenarios and business priorities according to the BCP. A comprehensive DRP includes specific information on hardware and software requirements for restoration, which systems and applications should be restored in what order, how to accomplish the restorations under multiple scenarios and how many user logins are required in given time frames. The scope of backups should address likely recovery needs and may include:

- Files
- Transactions
- OSs
- Databases
- Patches
- Configurations
- Access control lists
- Reports
- Applications

By definition, a disaster scenario is one under which normal operations are interrupted. Accordingly, a well-designed DRP will include at least a primary and alternate for every activity and should ideally be written in ways that allow a wide variety of staff members to complete assigned tasks. The plan should also be tested rigorously on a recurring basis (typically annually or every few years). The purpose of this testing is to uncover any weaknesses in planning or misalignments with evolving technology and provide the experience needed for team members to enact the plan effectively even under conditions of significant stress.

4.5 Data Life Cycle Management

Data is commonly viewed as following a six-phase lifecycle. Initially, data is created. It is then stored and thus becomes available for use, which is its third phase. In modern enterprises, data use frequently involves sharing, regarded as a fourth phase in a cyclical approach. When data is no longer needed for regular use, it is archived, and if its lack of use extends long enough, it may be destroyed.

In practice, this lifecycle contains many caveats. Creation may refer to original collection, but it may also be a product of synthesis from other data. In this way, the creation of data may result directly from sharing. Conversely, even in modern enterprises, not all data is shared. The archiving of data is a good practice, but it might be mandated by law, regulation or contract. On the other hand, law, regulation or contract might also prohibit archiving of data beyond strict necessity and instead require that it be verifiably destroyed by a certain date. **Figure 4.7** shows the data management life cycle.

CHAPTER 4— INFORMATION TECHNOLOGY AND SECURITY

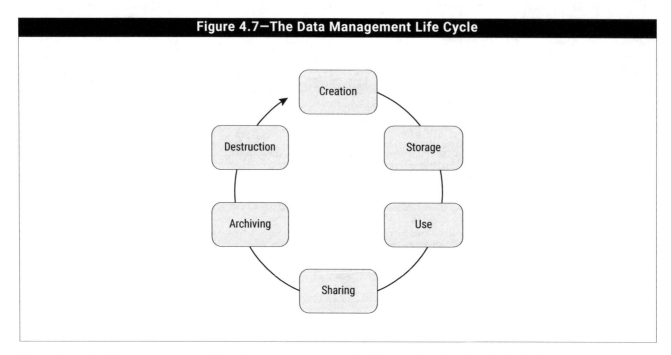

Figure 4.7—The Data Management Life Cycle

Throughout the data-management lifecycle, numerous areas of risk can arise. One is a lack of clear ownership of data, which makes it difficult to verify what requirements apply to the data and obtain approval to take steps such as archiving, sharing or destruction. The risk practitioner should review and assess the data ownership and management process of the organization, including the protection of data from improper disclosure, modification or deletion. Additionally, all staff who work with data in any form should be aware of the risk associated with improper data management (such as data leakage) and their obligation to comply with the policies and standards of the organization. These rules should be maintained and enforced.

IT management plays an active role in mitigating risk and supporting risk management activities. Above all, IT managers must abide by their own rules; this is crucial to developing a positive risk culture and promoting the integration of risk management principles into IT systems and projects. The risk practitioner should be observant of daily practices in the IT department and bring undocumented risk to the attention of accountable managers.

4.5.1 Data Management

Enterprises commonly regard data as being among their most valuable assets; however, attempting to protect all data in modern enterprises may create unsustainable cost or unacceptable risk. The goal of effective data management is to ensure that data is appropriately protected according to its value at all times and in all locations, including at rest, in transit, during processing and when displayed. The risk practitioner must also be aware that data protection applies to all formats, including:

- Paper printouts and notes (e.g., reports, scratch pads)
- Magnetic media (e.g., hard drives, universal serial bus [USB] storage)
- Optical drive media
- Audio and video broadcasts
- Photos and screensavers
- Discarded material, especially anything intended for recycling

CHAPTER 4— INFORMATION TECHNOLOGY AND SECURITY

Determining appropriate protections for data requires first identifying data in terms of its use and then classifying it based on its value, which is derived from the importance of the business processes that use it. Protection of data therefore starts with reception. At the point that data are collected, the purpose and the intent of collection tend to be clearly understood, and this is the opportune time to classify data on that basis. All data should be subject to validation checks before acceptance or processing to ensure appropriate formatting. In digital data collection, this includes ensuring that collected data strings do not contain embedded commands or other content that might adversely affect automated processing systems, such as structured query language (SQL) code. A rigorous process of data validation includes:

- Range checks (e.g., allowable data values)
- Format checks (e.g., configuration of data)
- Special character checks (e.g., no script commands)
- Size (e.g., data aligns with expectations for type and purpose)
- Likelihood (e.g., entries of correct form that are statistically unlikely and suggest an error)

Data validation may be done either with a whitelist of allowed data or a blacklist of prohibited data. A whitelist is the preferred approach in environments in which input data validation is based on static or infrequently changing values, where it is easier to anticipate what values would be reasonable than to anticipate what values might be malicious. The use of a common library for whitelisting can ensure that data validation occurs in a consistent manner across multiple applications.

Blacklists are useful in environments in which the range of valid values is extremely broad, but in which only a few known values should be prohibited (such as a dirty word search) and these are either entirely static or are updated rarely.

> Note: Blacklist design and maintenance can be particularly tedious whenever the same term may be expressed in multiple ways, a phenomenon known as "canonicalization." This is particularly common in web environments. For instance, the expression www.isaca.org can also be written as:
>
> - http:/www.isaca.org
> - http:/www.isaca.org/
> - http://isaca.org
> - https://www.isaca.org/Pages/default.aspx
> - http://www%2eisaca%2eorg
>
> Where there are many ways to input the same command, a blacklist that specifies only a subset of the possible forms may inadvertently accept an expression whose value is meant to be prohibited, leading to attacks such as directory traversal or bypassing firewall rules.

Data protection also requires assurances that changes made will not affect the integrity, precision or accuracy of the data or of data processing operations. Mechanisms that may assist in meeting this requirement include:

- Data checks and balances of input compared to output
- Checks of normal compared to abnormal levels of processing
- Anti-malware detection software
- SoD defined at every level of a system or application
- Processes requiring approval for transactions (all or based on thresholds)

One of the most important elements of data protection is control over the permissions and authorization levels of users that can access data and applications. The risk practitioner should be sure that users are granted the fewest

CHAPTER 4— INFORMATION TECHNOLOGY AND SECURITY

permissions that they need to perform their job functions and that these privileges are granted only for the time periods in which they are necessary, an approach to permissions management known as "least privilege." Maintaining a least-privilege environment requires regular review of user access permissions and the revocation of permissions whenever a user leaves the enterprise or changes job roles and is key to protecting data from improper alteration or disclosure.

Data commonly moves between networks and systems to serve multiple purposes within organizations. While in transit, data may be vulnerable to interception and unapproved removal (exfiltration). Enterprises can limit the risk of data transfer by sending data between network endpoints using encryption such as Transport Layer Security (TLS), which appears as an https: prefix in web addresses. Risk practitioners should be aware that some older browsers or systems may still be using Secure Socket Layer (SSL) for https: traffic, and that even the latest version of SSL is no longer considered secure. Support for SSL 3.0 can be disabled in many browsers to ensure use of TLS only.

Data should also be protected while at rest, which includes keeping sensitive data in separate networks or on systems that are not accessible to unauthorized personnel (isolation) through some combination of role-based and physical access controls. Effective data isolation for sensitive data may include network segmentation through the use of firewalls, virtual local area networks (VLANs) or other technologies that serve to limit access. Encryption at both the volume and file levels is also commonly available in modern information systems and should be used where practical. At a minimum, portable or removable devices should be configured with whole-disk encryption to limit the potential for data loss associated with loss or theft of physical media.

4.5.2 Data Loss Prevention

In addition to the techniques and methods discussed in the prior section, organizations concerned about safeguarding data frequently implement specialized software for data loss prevention (DLP). Full-scope DLP solutions leverage data classification schemes to first determine what controls should apply to data and then apply policies to how data can be accessed, moved, shared or stored based on its classification. Violations may result in automated alerts, mandatory encryption or other automated protective actions meant to safeguard the organization and prevent data loss.

DLP generally monitors and controls endpoint activities as well as reviewing data flows within enterprise boundaries. In environments where compliance reporting is important, such as those involving law and regulation specific to certain types of data, implementing DLP can facilitate reporting obligations. DLP can also be valuable in determining when and how data moves as a means of mapping business processes and understanding key points in production processing.

4.6 System Development Life Cycle

Complex systems can be significant sources of risk. Systems exist in all organizations, and it is common for specialized requirements to result in customized systems, including application software and integrated platforms for particular business use. Effective management of systemic risk is best accomplished through a rigorous and comprehensive system development life cycle (SDLC).

An SDLC describes project management as a series of phases that provides structure and audit-based accountability to a project. Although specific development methodologies may use unique terminology or variations in ordering, there are certain traits and activities common to most SDLC structures in use today as a result of their common purpose. One example of an SDLC is seen in **figure 4.8**.

SDLC Phase	Phase Characteristics	Support from Risk Management Activities
Phase 1—Initiation	The need for an IT system is expressed and endorsed by a sponsor, and the purpose and scope of the anticipated IT system is documented.	Identified risk existing within business processes or activities of the proposed system is used to develop system requirements, including security requirements and a security concept of operations (strategy).
Phase 2—Development or Acquisition	The IT system is designed, purchased, programmed, developed or otherwise constructed. This may be done within or outside the organization.	Risk identified during this phase supports security analyses of the IT system that may lead to architecture and design trade-offs during system development.
Phase 3—Implementation	The system security features are configured, enabled, tested and verified.	The risk-management process supports implementation against requirements and within its modeled operational environment. Decisions regarding identified risk are made and documented prior to system operation. In many environments, this process culminates in a formal authorization of the system for production use.
Phase 4—Operation or Maintenance	The system performs its functions, perhaps undergoing periodic updates or changes to hardware and software or otherwise altered to better align with changes to organizational processes, policies and procedures.	Risk management activities focus on maintaining acceptable levels of risk as established by the earlier authorization decision. Organizations may use periodic reassessment, or a continuous monitoring methodology as means of accomplishing this maintenance of risk alignment.
Phase 5—Disposal	The system is formally removed from production use, which may include disposition or reallocation of information, hardware and software. Depending on the business case for disposal, this phase may be triggered by migration to a replacement system or discontinuation of the processes driven or supported by the system.	Risk management in this phase focuses on ensuring that system migration is conducted in a secure and systematic manner, hardware and software are properly disposed of and residual data are handled appropriately.

Figure 4.8—Characteristics of the SDLC Phases

Source: ISACA, *CISM Review Manual 15th Edition*, USA, 2015, figure 2.25

The SDLC is fundamentally a methodology intended to support effective project management. There are numerous approaches to managing projects; however, many core principles are the same even where particular styles show considerable difference. Proper oversight, clear requirements, user involvement, communications between team members and users and regular review of project progress are all critical to project success under both waterfall and Agile approaches to project management. The risk practitioner will often find projects at risk of failure and should be able to identify the causes of these looming failures, recommend solutions and report the risk to management.

System development projects fail for many reasons, including:

- Changes in requirements arising from new business priorities, incomplete understanding of initial goals or poorly managed additions (scope creep)
- Non-availability of resources, which may be trained staff, budget, technology or something else on which success is dependent
- Delays in delivery of outsourced components
- Results of developed or outsourced technology that misalign with what was expected
- Excessive complexity that exceeds the capacity of management or monitoring
- Lack of leadership, accountability or oversight
- Unrecognized symptoms of failure
- Lack of coordination with suppliers

CHAPTER 4— INFORMATION TECHNOLOGY AND SECURITY

The risk practitioner should verify that project design, development and testing as to not solely focus on the business functions of the project. Security is ultimately a necessity in all modern business processes, and it is considerably easier and less expensive to address security requirements by building them into the project timeline at the onset than to attempt to add them into the project later. Security requirements should be specified, validated, and tested for utility in the same manner as other requirements to show that they reduce the risk associated with confidentiality, integrity and availability. The risk-management process should be incorporated into each step of the SDLC, and security requirements should be integrated into the project life cycle.

Key security tasks to perform during the SDLC to expressly document the risk associated with the development of a new program include:

- Security categorization of the proposed system—what are its confidentiality, integrity and availability requirements?
- What impact would an outage have on critical business processes (BIA)?
- Privacy impact assessment—what sensitive data are processed, stored or transmitted by this system? What laws or regulations apply?
- Vendor/supplier security or risk assessment—are key processes reliable and secure?
- Is there security training for all development staff and a secure environment? Are there secure code practices?
- What is the level of awareness of inherent vulnerabilities present in, or associated with selected technology or operational environments?

The ability to audit and review the applications or systems after implementation is also an important part of the SDLC. Maintenance is both inevitable and prone to changing assumptions and features associated with a complex system. Assessment and auditing of both changes and operating status are necessary to ensure ongoing proper operation of the system.

As the system moves through the phases of the SDLC, new risk may emerge, and previous risk assessments may need to be revised. This requires the risk practitioner to work with the project team to identify and determine optimal responses to risk, something that is especially important when changes are proposed to an existing application. Risk practitioners should be alert to whether the development team is following the standards and policies of the enterprise both for secure system development and the implementation of controls. Failure to adhere to these requirements may represent a risk to the enterprise and should be escalated to management for resolution wherever identified.

Additionally, the risk practitioner should be aware of the distinction between project risk, which deals with the consequences of failing to deliver the promised results on time and within budget, and risk that the deliverables may not meet the needs of the organization. The SDLC presented here generally reflects a Waterfall methodology in which requirements are finalized before development begins. In situations that have substantial delays between the preparation of detailed requirements and shipment of final deliverables, there is substantial risk that changing circumstances will make what is delivered irrelevant even if all deliverables are met. Agile development methods based on regular releases of minimum viable products and regular engagement with stakeholders for feedback on deliverables in progress can help ensure that projects focus on delivering what the organization needs.

4.7 Emerging Trends in Technology

New technologies continuously emerge on the modern market. Pressure to implement new technology is often influenced by inflated expectations of its utility and maturity and a focus on product functionality without attention paid to security. The desire to deploy new technologies for competitive advantage may cause the enterprise to lose sight of the business risk involved unless the risk practitioner is diligent and makes a compelling case for robust testing and assessment. One way to accomplish this is to use foresight and forecasting practices and techniques that have a longer-term outlook.

Emerging technologies often provide indicators years in advance of their potential. It is the job of the risk practitioner to consider the potential risk and controls for the application of these technologies that may present value to the enterprise. A well-managed change control process ensures that new technologies are not implemented until the security team has been able to validate the security impact of the change and enable appropriate controls. However, it is not good enough to wait until a technology has been implemented before actually reviewing and securing it. In many cases, if the IT department is unwilling to review and integrate new technologies, the business will find ways to implement them regardless (sometimes called Shadow IT), putting the organization in a vulnerable position where its true risk profile is unknown even to senior executives.

The risk practitioner should evaluate and assess the approach of the enterprise to accepting new technologies and the attitude of the security team and IT operations teams toward reviewing and securing the new technologies as they become available. The risk practitioner is considerably more likely to get the attention and support of senior management by leading the effort to demonstrate how new technologies can be safely incorporated into the enterprise than by adopting an obstructionist posture.

4.7.1 Omnipresent Connectivity

Computer networking goes back more than 50 years, and the Internet gained popularity in the 1990s. However, as recently as 2005, people connected on demand from fixed points tethered by cables and were otherwise disconnected. Improvements in wireless networking paired with miniaturization have reversed this situation: people today tend to be constantly connected through smartphones and other mobile devices that they carry with them. Whereas information sharing was once something done at specific times, people are now able to share information among themselves at any time.

The ability to connect has led to demonstrable changes in how people behave. Details that were once painstakingly memorized can now be consulted for quick reference as needed; as a result, people are less likely to gain mastery of specific skills in advance of direct need. Complex tasks can now be observed in video form rather than described in step-by-step paper manuals, so written instructions have become less precise. The intense focus that was the hallmark of analytical tasks for centuries has been replaced with a need to multitask and collaborate across disciplines.

For the risk practitioner, omnipresent connectivity means that it is harder to safeguard information, both because people are more able to share things and because they are more inclined to do so. It means that sensitive information might be recorded or photographed at almost any time—including by malware that users may not even be aware has been installed on their mobile devices. Under such circumstances, security awareness training takes on even greater importance (see section 4.9 Information Security Awareness Training).

Bring Your Own Devices (BYOD)

Omnipresent connectivity means that people are more accustomed than ever to using technology in their daily lives. In many cases, this personal technology is an efficiency multiplier that makes people more productive. When people choose their own technology, they also tend to prefer it over imposed options, and infringing on that freedom of choice can hurt morale.

In addition to convenience and preference, enterprises are also discovering that there may be direct cost benefits to allowing workers to bring their own devices to work. The average knowledge worker today has personal devices—a smartphone, a laptop, etc.—that meet or exceed the capabilities of typical procured equipment, something that would not have been true even two decades ago. People also tend to keep their technology up to date on a more frequent refresh cycle than enterprises typically embrace. All of these factors make BYOD an attractive option for enterprises that want to connect with workers in a more dynamic economy.

For the risk practitioner, BYOD should be considered a form of risk sharing, with emphasis that the consequences of a data breach or other security event remain primarily applicable to the enterprise and not the individual who owns the device. Users bringing their own devices to work should be subject to controls that manage this risk, such as allowing access to organizational data only through remote desktops that prevent transfer of actual data to user endpoints. Depending on the line of business, it may be that certain activities are best limited to organizational equipment. Specific risk appetite for BYOD will vary by enterprise, and that is to be expected. What is important is that risk is fully assessed and documented in a way that allows effective management. Simply renouncing corporate control over information flows is a likely recipe for disaster.

The Internet of Things

Traditionally, organizations have drawn a sharp line between IT, which deals with the processing of information, and operational technology (OT), which manages and controls physical processes. This line has recently blurred with the mass adoption of TCP/IP as the basis for networking between OT devices and the ability to embed into practical devices operating systems that have predominated in IT workspaces (such as Windows and Linux). The result is that more IT implementations are touching the physical world, while more OT implementations are taking information processing into account.

While numerous instances of OT remain disconnected or use proprietary communications methods, there is a growing number of practical devices built to leverage IT and communicate among themselves and with IT assets. This IoT encompasses everything from light bulbs to climate control systems to appliances. For home use, these IoT "smart" devices are typically touted as conveniences, such as allowing a resident to play particular music by verbally asking a smart "hub" to do so or receiving updates by text message that a refrigerator has been left open. In workspaces, IoT devices promise substantial cost reductions by allowing dynamic programming of thermostats, window shades and other energy-saving measures to align with employee work times.

IoT devices tend to prioritize functionality and may make little or no provision for security. Smart devices have been successfully hacked and used as sources for network traffic floods against particular endpoints, a technique called as distributed denial of service (DDoS). Devices programmed to respond to verbal cues are obvious targets for attacks that seek to record or pass along information that may be overheard, and climate controls that can be manipulated over the Internet could feasibly be altered in ways that greatly increased cost, made workspaces uncomfortable, or even caused physical damage.

Vendors respond to market pressure, and security measures have been added to some IoT devices in response to these concerns; however, the basic tension between operating efficiency and the cost of security is hard to overcome. Additionally, the functional tasks linked to IoT devices tend to be seen as part of the physical plant or facilities area of concern. Security processes that are commonplace in IT may not be part of the approach taken to designing IoT implementations from the outset. In many cases, the presence of these devices becomes known to information security staff only after they are already deployed, and enterprises find themselves reliant on compensating controls (see section 3.6 Control Types, Standards and Frameworks).

The risk practitioner should anticipate the appeal of smart devices and inquire about any IoT plans or implementations already in place. Where possible, the risk practitioner should bring together facilities and information security staff before smart devices are deployed and encourage the use of the SDLC for IoT projects. Building solutions with security in mind tends to be both cheaper and more effective than trying to secure a deployed system after the fact.

4.7.2 Massive Computing Power

As networking has expanded and gotten both faster and cheaper, it has become increasingly easy to connect multiple computers in ways that allow them to collaboratively process the same data in complex transactions. This connectivity advantage coincides with an increase capability on a per-processor basis, as multi-core processors have become standard and operating systems have gained the ability to leverage much larger amounts of memory. The combination of these factors, along with virtualization, is what makes cloud computing possible (see section 4.2.6 Virtualization and Cloud Computing).

For many enterprises, transitions to the cloud mean more easily predicted and controlled hardware utilization costs, smaller IT staffing requirements, and greater business flexibility. As previously discussed, there is also direct risk associated with moving to the cloud. However, the massive computing power assembled by cloud architectures has implications beyond assigning accountability in the event of disruptions.

Decryption

Organizations frequently rely on encryption (see section 4.8.6 Encryption) to safeguard their data should an unauthorized third-party gain physical access to a device or successfully authenticate to a user network. The assumption with encryption is that any files obtained will be unreadable without the correct key. Using computers to attempt to break encryption by finding the mathematical key to unlock it by brute force is nothing new. Until recently, it was infeasible that anyone outside a well-funded government agency would be able to assemble sufficient processing power to accomplish that task. The ability to rapidly provision computing power through cloud providers has eliminated this restriction, with pattern analysis and brute-force attacks now within the grasp of individuals or groups with modest means.

Quantum computing, based on principles of quantum mechanics that allow indicators to hold multiple binary states simultaneously, is also driving a revolution in computing capabilities. For some tasks, quantum computers appear to be no more effective than traditional processing, while they can perform others in dramatically reduced times. As the advantages of quantum computing become more widely understood, and prices fall, it can be expected that cloud providers will leverage these systems in ways that provide the most benefit within their overall architectures. Meanwhile, the aggregate computing power available within the cloud continues to grow.

For the risk practitioner, this means that it is difficult to predict how long an encrypted file might protect data if it were stolen. Encryption remains a vital part of information security, but it cannot be taken for granted that data is safe simply because it is encrypted.

Deepfakes

People tend to assume that they can trust what they see and hear. In the era of massive computing power, this is no longer the case. False audio and video can now be created using digitally manufactured imitations of a person based on samples. For political figures and celebrities with significant public profiles, obtaining these samples is astonishingly easy: many people express themselves with a wide variety of facial and body movements as well as verbal intonations and cues in any recorded presentation. Processing these inputs as a baseline, algorithms can reconstruct convincing replicas of human movement where one person is redrawn as another, while words and phrases are filtered in ways that make them sound like the target's voice.

It is possible to digitally detect these "deepfakes," but doing so takes considerable expertise. The typical person listening on the other end of a phone or watching a replayed video would be quite convinced by basic sensory recognition that the impersonator was who they claimed to be. This technology poses substantial challenges for organizations that have procedures to authorize financial transfers, release data or carry out other potentially sensitive tasks on verbal or visual approval of an authority figure.

CHAPTER 4 — INFORMATION TECHNOLOGY AND SECURITY

Risk practitioners operating in high-risk environments can guard against deepfakes misdirecting operations through a combination of security awareness and administrative controls. For instance, requiring that a particular transaction not only have the verbal approval of the CEO but also written approval from another executive, or using a dedicated callback number to obtain confirmation after receiving initial outreach, can help inoculate an organization against this kind of technical deception.

Unfortunately, no such protective measures are sufficient to guard against impacts to public opinion that may arise from falsified video or audio recordings, particularly when dealing with enterprises or issues subject to substantial public tensions. Many people for whom it is easiest to gather the data points needed to produce deepfakes are in positions where reputational risk is significant. Until there is an easy way to identify deepfakes and a mutually trusted third party to do so, it can be expected that deception for political or commercial gain will remain an emerging trend requiring a robust risk response.

Big Data

Proliferation of devices and computing power in an environment of omnipresent connectivity has led to an enormous expansion in the volume of data—by some estimates, more than 1.5 megabytes per second created for every human on Earth. Just a decade ago, actually storing data in such quantities would have been impossible. Since then, however, storage costs have fallen dramatically with the advent of cloud computing, and increased connectivity has made accumulation of data an everyday occurrence. Analysis of these vast amounts of data has in turn shifted from delivering competitive advantage to being a necessary precondition of business for many organizations.

For all the benefits that big data has to offer, data-centric enterprises also face risk. Accumulation of substantial volumes of data has implications for privacy, which is perhaps the most unpredictable and dynamic field of law and regulation associated with cybersecurity (see 4.10). Enterprises that have data piling up on a constant basis also struggle to keep it from becoming disorganized "dark" data that resists correlation and analysis. Finding the right experts to design effective data analytics is a growing concern, and retaining data beyond its intended use can result in reputational risk as well as fines or penalties for breach of contract under conditions where collection was explicitly limited to particular periods.

4.7.3 Blockchain

Data has been organized and stored in databases for decades. Traditionally, databases have been stored as tables and accessed either hierarchically or relationally. Recently, the capability has emerged to store data in blocks that are chained together, maintaining a timeline of transactions that allows the history of data to be determined without centralized processing. The absence of centralization facilitates extraordinary transparency in transaction logging that is inherent to the blockchain structure. The chain structure also makes it very difficult to falsify a blockchain entry, because each entry contains its own digital hash as well as the hash of the block before it (see 4.8.6 Encryption for more information).

Blockchain has been implemented in pseudo-currencies based on cryptography (e.g. Bitcoin, Ethereum). However, its potential implications extend beyond currency. Blockchain has been touted as a way to revolutionize banking, healthcare, property titles, and even voting. Risk practitioners in industries affected by blockchain should take care to remain apprised of what developments are underway.

4.7.4 Artificial Intelligence

For decades, scientists have attempted to create in computers the ability to behave autonomously in a manner that mimics human intelligence. Precisely what this means has long been subject to debate; as machines become increasingly capable of processing logical decision matrices, the parameters for intelligence tend to exclude what is

CHAPTER 4— INFORMATION TECHNOLOGY AND SECURITY

accomplished, so that artificial intelligence (AI) is sometimes seen as *by definition* an unattainable standard. One approach, suggested by computer scientist Alan Turing in 1950, is that a computer could be called intelligent when it was able to engage in behavior indistinguishable from a human. This approach has been both praised and criticized, with the principal criticism being that observed behavior does not demonstrate intent.

Regardless of where one draws the line between imitation and intelligence, computers today are capable of accomplishing tasks that mimic human behavior within narrow constraints. Digital assistants such as Apple SIRI® or Windows Cortana® are already capable of answering questions based on queries that leverage omnipresent connectivity. Device hubs such as Amazon Alexa® and Google Home® have extended this to manipulation of IoT devices, and considerable attention is being given to efforts that would apply this to the physical world in substantial ways, such as the provisioning of self-driving cars.

Risk practitioners involved with AI should keep in mind that there can be distinctions between human and rational response, and that any rules that can be positively verified can likely be exploited by those with intent and awareness. Even without intentional manipulation, there is considerable divergence between human and algorithmic goals. Decision trees and use cases should be subject to testing commensurate with the potential impact associated with the threat of misinterpretation or deception.

Oversight of AI is especially important in cases involving machine learning (ML), algorithms designed to refine themselves over time based on results. Humans know that habits are highly useful and make it possible for us to do things that would otherwise require far too much effort; however, bad habits may be very difficult to break. The same basic dynamic applies to ML systems. Where "improvement" is poorly or incorrectly defined, iteration aimed at refining outcomes may result in lasting deviation from goals. Heuristic systems used in intrusion detection and prevention are particular areas of concern for ML, where an errant determination that malicious traffic is normal could result in willful blindness to an intrusion underway.

Part B: Information Security Principles

One goal of risk management is ensuring that technology used in the enterprise is adequately protected, secure and reliable. As with IT, the risk practitioner need not be an information security expert but should be sufficiently knowledgeable as to ensure that the risk assessment and response program evaluates new technology and provides effective advice on how to deploy and use it within acceptable risk boundaries wherever possible. Some risk-based considerations regarding the deployment of new technology include:

- Training for users and administrators
- Creation of policies and procedures
- Inclusion of systems in backup schemes and continuity plans
- Assignment of risk ownership
- Consent of information owners for any technology that may handle sensitive information
- Review of legal or regulatory requirements
- Assignment of responsibility for monitoring and reporting on proper technology use

Older systems, commonly called "legacy" systems, frequently require special attention because of gaps between their designs and current security standards. Legacy systems might lack security features entirely or have features that are misaligned with observed threats. Where systems are reaching the end of their operational life, they may also be more susceptible to failure as a result of aging. The risk response options for legacy systems tend to be limited due to the cost involved in replacing or upgrading them. Some legacy systems may not be replaceable at all—a common occurrence for operational technology in industrial environments. Under such circumstances, the risk practitioner should bear in mind that his or her role is not to reject a particular course of action but rather to work with the risk owner to create an acceptable level of risk. Compensating controls—such as additional backups, spare parts, strong network segmentation, cross-training, documentation and increased monitoring of systems performance—may be effective supplements to existing security where conventional improvements are either technically infeasible or cost prohibitive.

Every system is the responsibility of a system owner. The owner of the system is usually a senior manager in the department for which the system was built. For example, the owner of financial data and the financial systems of an enterprise might be the comptroller or chief financial officer (CFO). The system owner is responsible for the proper use and operation of the system and usually must approve expenses for system implementation, changes, upgrades and removal. Ownership does not mean literal responsibility for all actions taken. Typically, system owners engage the IT department or an external supplier to manage and operate systems under their ownership on behalf of their functional areas. However, transference of responsibility does not mean transference of accountability, except in rare cases in which a supplier might accept full legal accountability for damages and losses. Responsibility can be delegated—accountability cannot.

Large enterprises may have many system owners from various departments, which can make it difficult to manage, oversee and ensure consistent operation of the systems in a coordinated fashion. For instance, an information system that supports the sales department may operate in a considerably different cultural environment than an information system that supports the finance department, and this difference in culture may affect the way the system is managed and protected. If the protection for each system is the sole responsibility of the owner of that system, then significant differences can exist in how the security and risk for each system is enforced. This is a substantial problem in modern networked enterprises, because information sharing and system dependencies make it likely that a breach or vulnerability in any one system is a risk to the entire enterprises. For this reason, an organization should take an enterprise approach to security to ensure consistent, reliable and secure operations. Two common methods of providing enterprise consistently are change control and system authorization. Asset inventory and configuration management are also important parts of an overall information security strategy.

CHAPTER 4— INFORMATION TECHNOLOGY AND SECURITY

4.8 Information Security Concepts, Frameworks and Standards

IT risk is often linked to information security, which is the protecting of information and information systems (including smart technology) from risk events. Information security controls are based on risk, and risk is the primary justification used to support information security activities.

Some enterprises have a mature risk-management process that is able to quantify risk with a high level of accuracy. In other enterprises, the term "risk" may be poorly understood and represent an emotional value based on perception. Risk is often difficult to measure, and because it is based on concepts such as likelihood and impact, it can be hard to quantify. Often, the predicted impact of an undesired event occur results in only direct, immediate effects, whereas actual events may have repercussions that affect tangential aspects of the business or are discovered long after the event has occurred (such as damage to reputation).

4.8.1 Likelihood and Impact

Chapter 2 IT Risk Assessment describes likelihood (also called probability) as the measure of frequency of which an event may occur, which depends on whether there is a potential source for the event (threat) and the extent to which the particular type of event can affect its target (vulnerability) after considering any controls or countermeasures that the enterprise has put in place. In the context of risk identification, likelihood reflects the number of events that may occur within a given time period (often an annual basis). **Figure 4.9** describes specific factors that can affect likelihood.

Figure 4.9—Factors That Can Affect Likelihood	
Factor	Volatility
Volatility	Unpredictability, also referred to as dynamic range; the degree to which conditions vary from one moment to another, making projections difficult
Velocity	Speed of onset, a measure of how much prior warning and preparation time an enterprise may have between the event's occurrence and impact, which itself can be split into speed of reaction and speed of recovery
Proximity	The time from the event occurring and the impact on the enterprise
Interdependency	The degree to which materialization of two or more types of risk might impact the enterprise differently, depending on whether the events occur simultaneously or consecutively
Motivation	In cases involving an active/sentient threat, the extent to which the perpetrator of the threat wants to succeed, which may result in a higher chance of success
Skill	The ability brought to bear by the perpetrator of an active/sentient threat relative to other perpetrators
Visibility	The extent to which a vulnerability is known which can make it a more likely target of attack

In addition to likelihood, risk also depends on the impact arising from a successful threat event. The risk practitioner should consider two forms of impact:

- Impact due to the loss or compromise of information
- Impact due to the loss or compromise of an information system

4.8.2 CIA Triad

The risk practitioner may find it helpful to evaluate an information system or data source on the basis of confidentiality, integrity and availability (CIA), the three principles from which all other characteristics of information security are derived. By considering information and systems according to this model, it becomes easier to identify whether and how a compromise or failure in one or more of these areas would affect the enterprise. For purposes of risk identification, simply noting that an impact may occur is sufficient; during risk assessment, the practitioner draws on these findings to develop a more detailed analysis that includes the particular extent of impact from each area of compromise.

It is important for the risk practitioner to have a solid understanding of what is meant by CIA, particularly the interrelationship between the three principles as well as nonrepudiation, a fourth principle that is derived from them. There are typically trade-offs between the three principles in any system making it impossible to simultaneously increase one of them without either decreasing at least one of the others or substantially increasing cost. For instance, increasing confidentiality typically increases processing time, which reduces availability. The CIA triad shown in **figure 4.10** illustrates this relationship.

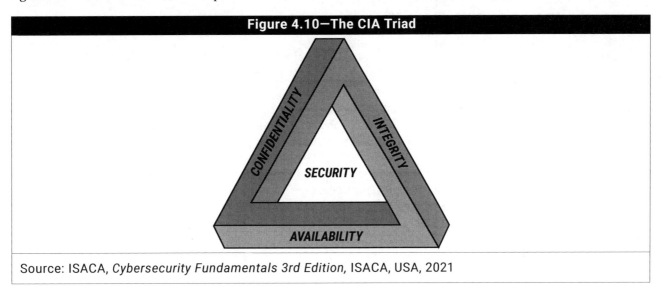

Source: ISACA, *Cybersecurity Fundamentals 3rd Edition*, ISACA, USA, 2021

Confidentiality

Confidentiality refers to the secrecy and privacy of data. Accordingly, a breach of confidentiality means the improper release of information, such as disclosure to an internal or external recipient not authorized to access it. Confidentiality is often discussed in terms of the need to protect classified military secrets or other categories of sensitive information, such as personally identifiable information (PII), personal health information (PHI) or intellectual property, and other specific types of data protected by law or regulation. However, the basic principle of confidentiality is to ensure that disclosure does not occur to any recipient not authorized access by the information owner.

In identifying risk related to confidentiality, the risk practitioner should look for policies or behaviors that violate either or both of the following principles:

- **Need to know**—individuals should be given access only to information that is needed in order for them to perform their job functions. For example, most users do not need to see credit card numbers; therefore, the numbers are masked or hidden except for the last few digits, which protects the cardholder data from being compromised by a person that did not need to see the entire credit card number.
- **Least privilege**—the level of data access afforded individuals or processes should be the minimum needed to perform their job functions. For instance, someone who needs to review data should not be able to edit or delete it.

Policies, practices or behavior that does not reflect these principles are a warning flag to the risk practitioner of possible risk related to confidentiality.

CHAPTER 4— INFORMATION TECHNOLOGY AND SECURITY

Integrity

Integrity refers to protection against improper modification, exclusion or destruction of information and applies to actions taken by both authorized and unauthorized users as well as processes or activities operating on the system. Whenever data are changed in a manner other than that intended by the data owner, integrity is compromised.

Maintaining integrity is typically a rigorous process that relies upon multiple levels of error checking and verification that may be difficult for the risk practitioner to fully evaluate without technical expertise. The principle of least privilege is also critical when identifying risk related to integrity, particularly when it comes to access by system processes and other memory-resident applications. Whenever a user or process has the power to change or delete data, the ways in which it interacts with data must be carefully considered to determine whether there is an integrity risk.

Availability

Availability refers to providing timely and reliable access to information. Timely access to data is often important to business processes, and in some cases, such as industrial control systems operating machinery or regulating power generation, near-real-time availability may be essential for safety and proper system operations.

When asking what levels of availability business processes require, risk practitioners should anticipate an initial answer from process owners that systems and data are essential and require 100 percent availability. In most cases, this is not accurate, as rapidly becomes clear when the financial costs associated with guaranteeing 100 percent uptime are presented. From there, the risk practitioner should expect to work with the business to determine the true availability requirements, which tend to be driven by what the organization is willing to pay to have higher availability. One way to identify potential availability risk is to compare current levels of availability with required levels; where there is a gap, there is a risk.

Nonrepudiation

In information systems, activity is often either anonymous or only weakly traceable to a particular individual or process. Nonrepudiation refers to a positive guarantee that a given action was carried out by a given individual or process and is an important part of tracing responsibility and enforcing accountability. Digital signatures and certificate-based authentication in a public key infrastructure (PKI) are examples of controls whose proper implementation and use can provide nonrepudiation, while shared or generic logon credentials do not provide the relatively weak assurance afforded by individual usernames and passwords. For purposes of risk identification, the risk practitioner should look for evidence of nonrepudiation in situations in which actions may have significant impact on an enterprise, such as approval of production code, deletion of records or disbursement of funds.

System Authorization

Many enterprises use a form of objective assessment and formal acceptance of risk associated with the installation and operation of information systems, culminating in the explicit authorization of a system to operate prior to it being allowed to do so. Previously known as certification and accreditation (C&A), this process is more commonly called security assessment and authorization (A&A) in contemporary language because of the use of that phrase in US National Institutes of Standards and Technology (NIST) special publications relating to risk management.[8]

During assessment, an information system is thoroughly reviewed to determine the security of its design, development, testing, deployment and operations. The evaluator examines the technical and nontechnical aspects of

---

[8] NIST is an agency of the United States government; however, many other countries reference NIST publications and frameworks or base their own frameworks on them, giving NIST outsize influence over terminology.

system operations to ensure that the risk associated with the system has been identified and to document any mitigating controls that may be in place. This means evaluating assessments, system documentation and the results of technical tests to verify that the system will operate within an acceptable level of risk, so the evaluator should be a technical and process expert. The culmination of the assessment process is a report to a suitably empowered senior manager recommending whether to authorize operation of the system. It is a good practice for evaluation to be done in parallel with the SDLC to ensure that any unacceptable risk is identified promptly and addressed to system and project management.

Authorization is the official decision by the senior manager—sometimes called an authorizing official (AO)—to approve an information system for operational use. This decision is the explicit acceptance of the risk documented in the evaluator's report, and it may come with caveats. For instance, a system may be authorized to operate only under certain circumstances or in particular time frames or certain risk may be accepted only for a limited time on the basis of response activities that are already underway but have not yet been completed. Although the AO typically concurs with the assessment recommendation, the decision to accept risk does ultimately lie with the senior manager, not the evaluator who performed the assessment.

Upon receipt of authorization, the system owner may operate the system according to the restrictions and for the time period granted in the authorization. Depending on the maturity of the organization and its processes, this may be a finite time frame—generally one to three years, with the caveat that any substantial changes in the system or the organizational risk profile require a new evaluation and reauthorization of the system—or indefinite under a system of continuous monitoring that keeps risk aligned with acceptable levels.

4.8.3 Segregation of Duties

Errors and intentional malicious acts are more difficult to detect when single individuals are able to complete potentially risky processes on their own. SoD is a basic internal control that prevents or detects errors and irregularities by assigning separate individuals responsibility for initiating and recording transactions and for the custody of assets. Most often, this takes the form of requiring one person to review and approve the work of another person. For example, a request for a large payment may need to be approved before it is processed, or a developer may need to submit code to a testing team to have it moved into production. SoD may also mean requiring two people to participate in a task simultaneously (often called dual control), a system used to control release of nuclear weapons.

The key to SoD is mutual exclusivity, meaning that a single person cannot execute both parts of the same transaction. SoD does not guarantee security, because it can be circumvented when all participants in the process agree to work together to bypass the intended effect of the control (collusion). However, because it is difficult to establish collusion, and especially challenging to maintain it over time, effective SoD increases the odds that an incorrect or fraudulent transaction will be detected and addressed.

4.8.4 Cross-training and Job Rotation

Enterprises can reduce their reliance on key staff by training multiple people with the same skills, making it possible for any of a number of individuals to step in to fill a vital role as needed. In many cases, this takes the form of cross-training, in which people on the same team are trained in one another's roles. Cross-training makes it possible for an enterprise to rotate people between different jobs.

Although job rotation is typically associated with business continuity and enterprise resiliency, it is also an effective security mechanism because it changes who is involved in which roles at which times, reducing the potential for collusion and increasing the odds that prior collusion can be detected by someone who is not part of the scheme. In environments where there is substantial concern regarding the potential for collusion, job rotation may take the form of mandatory vacations in which people are removed from their regular duties specifically to allow independent verification.

One risk associated with cross-training is that employees with substantially broader skill sets may be more attractive to other employers. The rotation of personnel may also result in decreased efficiency during times of transition, and employees who know that they are about to move out of jobs may be less diligent than those who expect to remain in them. Nonetheless, in high-risk environments, this potential loss of productivity may be warranted. This is particularly common in financial institutions and positions that deal with moving large amounts of money.

4.8.5 Access Control

Managing access to information systems and data is one of the most challenging aspects of information security. Access control is commonly addressed through the concepts of identification, authentication, authorization and accountability (IAAA).

Identification

With a unique identifier assigned to every individual, device or process that has access to a system, it is possible to track and log activity on a per-user basis, creating the possibility of targeted investigation were a problem to arise. Identification is often provided through a user ID, customer account number, employee identification number or other unique element. Sharing of user IDs should be prohibited, and risk practitioners should verify that the process of issuing an employee ID is reasonably secure and requires proper authorization.

Authentication

Authentication is the process of validating an identity once it is presented. The purpose of authentication is to ensure that one person cannot spoof an identity or masquerade as another user, which includes prevention of ID sharing by more than one person. Authentication is done using one or more of three factors:

- Something you know (e.g., user ID, password)
- Something you have (e.g., token, authenticator)
- Something you are (e.g., fingerprint, iris scan)

Authentication by knowledge means presenting a password, code phrase or other secret value to validate identity. Complex knowledge questions may be extremely personal, include artistic or visual representations, exercise particular skill sets in certain ways or otherwise map to something that arbitrary individuals would be unlikely to know. However, knowledge is often subject to replay attacks, even when what is asked is complex—sometimes more so, because complex passwords or knowledge responses tend to be used for an extended period of time. Time is the enemy of knowledge authentication, because any knowledge question can eventually be figured out, allowing one person to log in as another. Therefore, changing passwords on a regular basis is recommended even in high-complexity environments.

Authentication by possession relies on some sort of item, such as a smart card, token or ID badge. Direct possession may be sufficient, or the user may generate a one-time password from a token using a pre-shared algorithm that cannot be guessed. Authentication based on possession was once the subject of considerable expense when installing the infrastructure, issuing the cards or tokens and operating and maintaining the system. Widespread ownership of mobile phones capable of running complex applications has dramatically reduced the cost, allowing possession-based authentication to become relatively commonplace. The ability to digitally lock mobile phones also mitigates a longstanding concern that a lost token might be used by an imposter if it has not been reported as lost or stolen. However, one downside remains: should the means of authentication be lost, stolen or damaged, the legitimate user may be denied access without easy recourse. Where availability is a primary concern, the risk practitioner should take care to examine what workarounds may be in place, keeping in mind that any workaround also represents a potential area of vulnerability.

Characteristic authentication relies on biometrics, factors present in physiology (e.g., fingerprints, iris scan, palm scan, etc.) or behavior (e.g., voice print, signature dynamics). Like possession, access to certain basic biometrics that were once expensive has been made attainable through the distribution of mobile phones, which may be able to unlock functions based on facial recognition or other fingerprint analysis. Purpose-built systems can even monitor body movement and other indicators to cut off access should there be indications that a device has changed hands, and it is generally infeasible to lose characteristics in ways that would deny access. However, some users find biometrics to be intrusive and may be resistant to it. In addition, biometric data may be subject to laws and regulations governing privacy, with potentially severe penalties in some jurisdictions should the data be compromised.

Authentication of devices or network nodes may also use characteristics as determining factors. Node authentication is used when limiting access to certain devices through the identification of IP addresses, MAC addresses or serial numbers of devices that are attempting to log on. These characteristics avoid the concerns of human biometrics because privacy is not a concern for devices; however, it may be possible to falsify (spoof) digital characteristics. Where positive control is needed for network endpoints or appliances, use of certificates can help positively establish identity and broker authentication using protocols such as 802.1x.

Multifactor Authentication

Single factors of authentication are rarely considered strong enough on their own to safeguard modern information systems. By combining more than one type of authentication during the login process, organizations can establish strong authentication that is difficult to breach in a reasonable time. This could include using a password with a smart card, a biometric indicator paired with a smartphone-generated code, or a combination of inputs from all three factors. Such arrangements are commonly referred to as multifactor authentication (MFA).

The most common MFA implementations in use today combine traditional passwords with codes sent to or generated by applications on personal smartphones. Risk practitioners should be aware that these arrangements, while convenient, make authentication reliant on access to devices that are outside of organizational control. Should a mobile device required for MFA be unavailable, the result is likely to be a temporary inability of a valid user to authenticate to organizational information systems.And, again, smartphones are among the most easily damaged, stolen and lost items in individual possession.

Authorization

After a user has successfully authenticated, the system must provide that user with appropriate levels of access. Authorization refers to the privileges or permissions the person will have, which may include read-only, write-only, read/write, create, update, delete or full control over data. Permissions may be applied on a per-file or per-folder basis. Proper employment of least privilege is important to ensure that employees only have the level of access they require to perform their job functions.

Isolation

Authorization is typically granted only for a limited period, which may be a fixed duration before timeout or a set period of time that the permissions are required. For instance, access to a building, network or application might only be allowed during normal working hours, and outside those hours, access is not allowed even with a valid account. Similarly, where location tracing places the source of an authentication request outside a likely geographic area, additional verification may be required, or the access attempt might be blocked entirely. This isolation on a time- or location-delineated basis prevents a user from using account privileges at times when they should not be required and allows an organization to maximize the effectiveness of its controls by reducing the potential scope of monitoring.

Accountability

The final part of the IAAA model is accountability. Typically accomplished through auditing, this action relies on systems logging or recording activity on a system in a manner that indicates the user IDs responsible for the activity. The effectiveness of accountability measures relies on both unique user IDs and effective authentication methods as well as ensuring that the logs are protected from modification or alteration. Audit logs should not be changeable by operations staff—even administrators—in order to prevent a user from disabling the logging function or altering log data that implicates their own account in suspicious or malicious activity. Because logs may contain sensitive data, even view access to log content should be strictly controlled.

4.8.6 Encryption

Encryption is a mathematical means of altering data from a readable form (plaintext or cleartext) into an unreadable form (ciphertext) in a manner that can be reversed by someone who has access to the appropriate numeric value (key). The foremost use of encryption is to protect the confidentiality of data by making data unreadable to anyone who is unauthorized. The science of creating encryption is called cryptography, while the broader study of creating and solving cryptographic puzzles is called cryptology.

In addition to confidentiality, novel uses of data encryption can also provide several additional risk management benefits, including:

- Integrity
- Proof of origin (nonrepudiation)
- Access control
- Authentication

Mathematically, encryption comes in two basic forms—symmetric and asymmetric—each of which has its own strengths and weaknesses, and these can be used in a variety of ways. The risk practitioner should be familiar with these concepts of encryption, which are central components of data protection (**figure 1.11**).

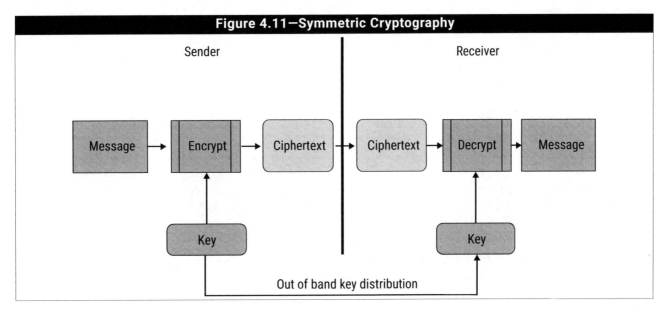

Figure 4.11—Symmetric Cryptography

The most common symmetric key cryptographic system currently in use is the Advanced Encryption Standard (AES), a public algorithm known as a block cipher because it operates on plaintext in blocks (strings or groups of data values). AES supports keys from 128 bits to 256 bits in size and is found a wide variety of encryption applications, including file transfer, content presentation, and the creation of VPNs.

Because the length of a key determines the number of possible mathematical combinations, shorter keys pose an inherent security problem as processing power increases. Where shorter keys are used, the entire key space might be tested against ciphertext by large computer systems within a relatively short period of time, via a technique aptly known as brute force. The effectiveness of this technique has expanded rapidly with the advent of cloud computing, which allows provisioning of massive computing power on a short-term basis. AES includes a number of advanced mathematical techniques that make its ciphertext stronger than prior algorithms. Nonetheless, ongoing advances in computing power virtually guarantee that any algorithm in use will eventually be rendered obsolete. Additionally, risk practitioners should also be aware that even the strongest algorithm with the most robust keys is only as secure as its key management system.

Symmetric key cryptography has two major disadvantages. One is the need for one party to deliver keys to another with whom it would like to exchange data. Where communications are established on a point-to-point basis between trusted parties, this may be workable through a variety of physical (out of band) methods, after which an existing secure channel can be used to send updated keying information on a periodic basis. However, there are numerous environments (such as e-commerce) in which customers are unknown and not trusted entities, making brokering a connection more difficult. Such arrangements are also at odds with the need for dynamic connections lasting only a short time.

Once such a channel is established, there are limits to what it can accomplish. For instance, if the same symmetric key is shared to all participants on the channel, there is no way to be sure which participant in a given key network originated a particular message, because everyone is using the same key. However, the alternative is to have a distinct key shared between every pair of participants, which creates exponential growth in key pairs for every added recipient.

Asymmetric Algorithms

Asymmetric key cryptographic systems evolved to address the issues that arise with symmetric key cryptosystems. The intent of an asymmetric algorithm is to provide a way by which a secure channel can be created between parties that lack prior awareness of one another.

In 1976, Whitfield Diffie and Martin Hellman published the first public example of encryption based on two different keys, a technique that has since been known as the Diffie-Hellman model. In a Diffie-Hellman setup, two mathematically related keys are created: a private key and a public key. It is computationally infeasible to determine the value of one key from the other, but the manner of their generation means that each has the effect, when used in an appropriate asymmetric algorithm, of creating ciphertext that only the other has the ability to return to plaintext. As a result of this relationship, the public key may be freely distributed and used as the means of encrypting any message that should be readable only by its creator, who has sole access to the private key that allows the messages to be decrypted. Additionally, the only messages that the public key can decrypt are those that were encrypted by the private key, to which its creator has sole access, so any recipient can be certain that the key creator was the true author.

Because asymmetric algorithms involve sharing the public keys of a public/private key pair, they are commonly referred to as "public key" algorithms. Each party using public key cryptography only needs one pair of keys, and the ability to freely share the public key overcomes the challenges of scalability seen in symmetric key cryptography.

Asymmetric algorithms are computationally intensive and slow relative to symmetric algorithms. For that reason, asymmetric cryptography is typically used only to encrypt short messages. The most common use of asymmetric algorithms is to distribute symmetric keys that can then be used by the participants for fast, secure communication, as seen in **figure 4.12**.

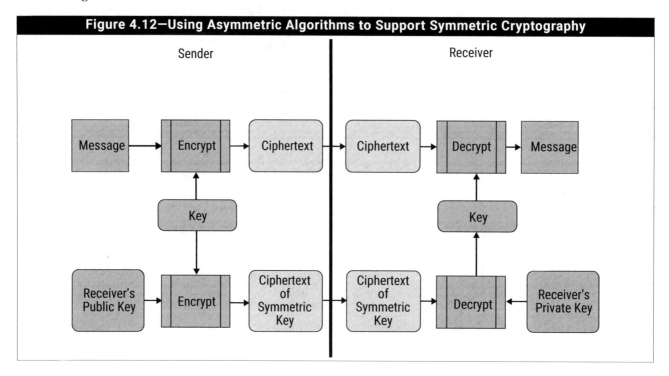

Message Integrity and Hashing Algorithms

Early computer networks used voice-grade telephone cable with limited bandwidth. The combination of slow speed and interference meant that error correcting was needed to ensure that what was received matched what was sent.

Parity bits, checksums and cyclic redundancy checks (CRC) were effective in detecting accidental errors, but these mechanisms added data to each transmission. To reduce both the amount of data and the computational power needed for error checking, cryptologists developed the concept of hashing, which is the mathematical transformation of data using an algorithm whose result is predictable, repeatable and entirely dependent upon the content of the message and of a fixed length (regardless of the length of the original message).

For any message provided, a hash algorithm calculates a fixed-length value commonly known as a digest, fingerprint or thumbprint. The length depends on the hash algorithm used; for example, SHA1 generates a digest of 160 bits, while SHA512 generates a digest of 512 bits. When a sender wants to ensure a message is not affected by noise or network problems, he or she can hash the message and send the digest along with the message to the receiver. The receiver can then hash the received message and verify that the resulting digest matches the digest sent with the message, as shown in **figure 4.13**.

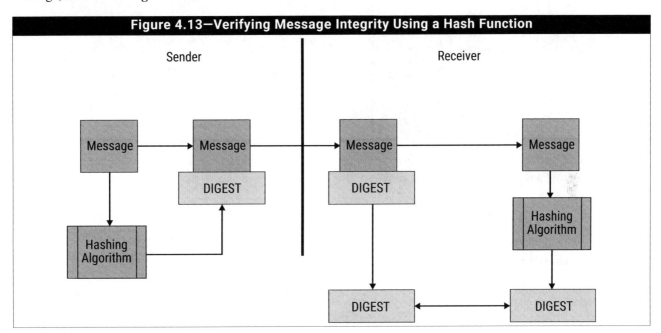

Hashing by itself provides effective protection against accidental changes in a message, such as those caused by interference, and this was its original purpose. However, a malicious user who intercepts a transmission that includes a hash will not be deterred by hashing, because they can create a new digest after altering the message. To guarantee integrity against intentional as well as accidental changes, cryptologists extended the basic idea of hashing to include features of public key encryption, resulting in digital signatures.

Digital Signatures

A digital signature combines a hash function with the ability of public key encryption to prove the author's identity. The hash function allows the receiver to verify that the message was not changed since the digest was added, because the hash of the received message is the same as the digest attached to the message. The receiver also knows that the digest came from the original sender, who has encrypted it with his or her private key. To access the digest, the recipient's ability to decrypt the digital signature with the sender's public key means that it must have been encrypted with the sender's private key and is, therefore, the genuine digest for the message as originally written. Moreover, the sender cannot plausibly deny it, because *only* the sender's private key could have encrypted it. This combination of verifiability and nonrepudiation makes private-key encrypted digests behave comparably to physical signatures, which is why the process of generating them is known as digitally signing messages.

Figure 4.14 illustrates the use of digital signatures to verify integrity and proof of origin.

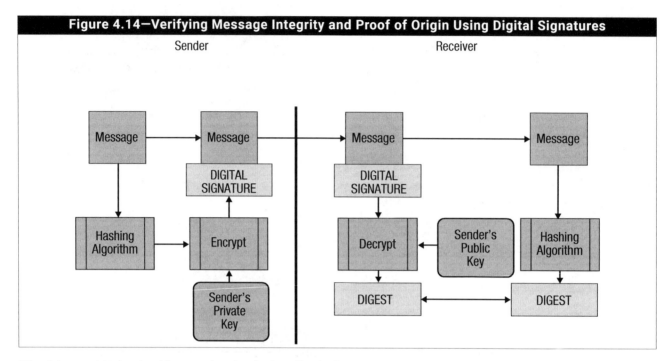

Figure 4.14—Verifying Message Integrity and Proof of Origin Using Digital Signatures

The risk practitioner should remember that just as physically signing a document does not keep others from reading it, digitally signing a message applies encryption only to the digest and does not make the message content confidential. A message may be:

- Encrypted and not signed (confidential but lacking integrity or nonrepudiation)
- Signed and not encrypted (not confidential but with integrity and nonrepudiation)
- Signed and encrypted (confidential with integrity and nonrepudiation)

Only a message that is both signed and encrypted is protected against unauthorized viewing, guaranteed to have arrived in the form that it was originally sent, and known to have originated from the person who claims to have sent it.

Certificates

Public key encryption ensures that the holder of a private key can decrypt what is encrypted with the corresponding public key. However, it does not prove that the person who owns the public key being distributed is who he or she claims to be. Certificates link public keys with specific owners by relying on the endorsement of a trusted third party known as a certificate authority (CA). The CA verifies identity by other means and generates the certificate on behalf of the owner of a public key so that the owner can prove that the public key belongs to them. By relying on this trust, the recipient of a message digitally signed with that public key of the certificate knows that the message was signed and sent by the claimed sender. The same person also knows that using the public key to encrypt a response will yield a message that only the intended recipient can open.

The format of a certificate is based on a standard called X.509, which ensures that certificates can be accessed by most Web browsers, systems, and software even if they are issued by different CAs. Each certificate is valid for a defined period of time (typically one year from the date of issue). However, the owner of a certificate may cancel it at any time during that period by notifying the CA, which will put the certificate on a certificate revocation list (CRL). Requests to validate a certificate that is on a CRL result in notification that the certificate has been revoked, warning that the certificate should not be trusted as a means of verifying identity.

Public Key Infrastructure

Public key infrastructure (PKI) refers to the overall implementation of certificates and the CAs needed to issue and verify public key cryptography. With a given PKI implementation, there is a top-level ("root") CA that manages all the certificates for users within the architecture. Through the root CA and any subordinate CA delegated by the root, any member of the group can gain access to the certificates of the other group members. This ensures that communication is secure and can only be accessed by the appropriate parties.

PKI implementations may be connected to one another and allow certificate verification using an external CA. When members of different PKI implementations are linked by relationships established between their respective CAs, a cross-certification agreement should be in place in order to provide a policy basis for this inter-connectivity. Additionally, the risk practitioner should ensure that mechanisms for certificate management are in place so certificates can be renewed prior to expiration. Once a certificate expires, results may include loss of access to encrypted messages and communications channels that depend on PKI-authenticated access.

Disadvantages of Encryption

Although encryption is extremely useful and is a central component of most data protection strategies, there are drawbacks to its use. One is that encryption requires processing, and its demands increase with the length of the key and strength of the algorithm. Encryption also increases network traffic based on the need to verify certificates, particularly in cases in which large numbers of small messages are being passed between distinct end points. The increased processing and transmission demands placed on IT infrastructures by encryption mean that it should be used only when necessary. Encryption is thus not a purely technical control but rather depends in part on policy at the organizational level. The need for policy extends to management of the CA, which can be used to issue credentials that will be generally accepted.

Encryption also presents a challenge to data availability. PKI implementations are typically robust to ensure that CAs and CRLs are always available, and there are emerging mechanisms for handling matters related to encryption keys that may provide even greater speed and reliability. Nonetheless, the risk practitioner should understand that encryption is not a fail-secure control: if keys are corrupted or lost, encrypted data will not be immediately recoverable and, in some cases, may not be recoverable at all.

Summary of the Core Concepts of Cryptography

Cryptography has many moving parts that work together in a complex series of operations. The following items summarize key information that the risk practitioner should know regarding cryptography:

- Symmetric encryption algorithms are excellent at providing confidentiality to large messages, but managing shared secret keys is difficult.
- Asymmetric algorithms are much slower than symmetric algorithms, but keys can be securely exchanged between users of asymmetric encryption using a well-established method.
- Because of the respective strengths of the two types of encryption, asymmetric algorithms are used to send symmetric keys, which are then used to encrypt messages.
- The two keys used by asymmetric algorithms (public and private) are mathematically related in a manner that lets each decrypt what the other encrypts, but it is infeasible to determine one key from the other, so public keys can be freely distributed while private keys are kept confidential.
- Hashing algorithms efficiently ensure message integrity against accidental errors by creating digests of fixed length that are mathematically linked to particular message values.

CHAPTER 4— INFORMATION TECHNOLOGY AND SECURITY

- Encrypting the hash of a message using a private key creates a digital signature that can be used to verify both message integrity and the origin of the message, which also means that the owner of the private key cannot deny sending it (nonrepudiation).
- Certificates provide assurance that the owner of a public key is who they claim to be. The assurance provided by a certificate depends on the degree to which the issuer of the certificate can be trusted.

4.9 Information Security Awareness Training

As technology has proliferated across societies around the world, the focus of information security has fundamentally shifted. The number of people whose workplace roles directly involves use of a computer or information system is orders of magnitude higher than was true just a few decades ago. People are also far more accustomed to using computers (including mobile devices), networks and data sharing services in their private lives than they were as recently as 2010. At the same time, there has been a realignment of computing resources from business to personal use. Whereas employees using computers at work once had access to systems vastly more powerful than anything usable in the home market, the average home computer is today as capable as the average work computer, while networked computing capabilities from cloud providers are available within cost ranges that even home businesses may be able to afford.

The familiarity with technology that people bring to their workplaces may reduce operation training costs and help boost productivity. It also creates risk. As barriers to entry into the computing workspace have fallen, so have barriers to becoming cyber threat actors. From "script kiddies" who mime the actions of professional penetration testers as a way of gaining experience to part-time dabblers in corporate espionage, all the way to foreign intelligence assets who have been recruited to launch cyber attacks from inside national borders, enterprises are faced with a bewildering threat landscape that includes more than a few operators who use basic home computing technology to carry out attacks.

Current threat actors have correctly identified human users as the weakest link in the information security foundation of modern enterprises. Security technology has improved dramatically; in contrast, human users are more vulnerable than ever. Employees accustomed to freely sharing information on social media and on their personal networks may also instinctively make information more widely available than authorized by organizational policy or seek workarounds to security features that they view as obstructing them from using computers in the manner that they would prefer to work.

Presented with realistic-looking emails or phone calls that claim to be from support or security services within the enterprise and asked to take some particular action immediately, many people are inclined to do as they are told. Worse, because of inadequate SoD and poor implementation of least privilege principles, their accounts often have the permissions within organizational systems to cause serious damage. By the time that these unwitting accomplices to cyber attacks become aware of what has happened, organizations may be facing permanent loss of data, damage to reputation, unauthorized transfers of funds or liability damages to third-parties whose data has been stolen.

To counter these threats, enterprises need to educate and train their workforce on proper tenets of information and cybersecurity. Primary is resistance to social engineering techniques by unauthorized individuals to get them to carry out tasks that benefit malicious agendas. The actual tasks tend to be simple, such as clicking a link, providing a phone number, or opening an attachment. In fact, these tasks fit into larger plans and strategies, potentially deploying malware, taking over users' accounts, or providing leverage to reach further into an enterprise without being detected.

Making human users more resilient to the tactics of threat actors is a major purpose of security awareness and training programs, but it is not the only purpose. Human users are also the first line of defense. Properly trained, human judgement can help identify attacks in progress even when directed elsewhere. By alerting response

CHAPTER 4— INFORMATION TECHNOLOGY AND SECURITY

capabilities within the organization, resilient people help create a resilient workforce, which in turn improves organizational information and cybersecurity.

Risk practitioners should review the scope of information security training and awareness programs against identified threats faced by the organization. At a minimum, the program should address these three things:

- Threats from social engineering, including deceptive emails (phishing) and calls or voicemails (vishing), which form the principal threat vector for attacks targeting human users
- Methods of alerting internal security response capabilities such as an IT service desk or incident response team to ensure prompt notification and response to any threat events
- Policies and regulatory requirements specific to the enterprise or its business sector

Leading enterprises also incorporate role-based training for administrators and others in roles that handle sensitive tasks or data as a way of mitigating role-specific risk. In addition, it is common to implement ethical phishing email exercises to assess how easy it is for employees to fall victim to a phishing scam. As a result, the enterprise can measure the level of exposed risk and the necessity of improving the awareness programs in place.

4.10 Data Privacy and Principles of Data Protection

Privacy and data protection are increasingly important to enterprises. The prominence of these areas of concern arises partly from competitive advantage, with customers expecting rigorous protection of their data. At the same time, numerous jurisdictions have established strong laws and regulations that pertain to the handling and protection of specific classes of data. Among the most impactful is the European Union General Data Protection Regulation (GDPR), which applies to data stored anywhere in the European Union and also sets limits on the transfer to other jurisdictions of covered data associated with EU citizens. GDPR establishes seven key principles:

- Lawfulness, fairness and transparency
- Purpose limitation
- Data minimization
- Accuracy
- Storage limitation
- Integrity and confidentiality
- Accountability

Many countries have their own approaches to privacy, and multinational organizations are subject to the requirements of any jurisdiction in which they do business. As penalties associated with non-compliance increase—for instance, violations of GDPR can extend as high as 20 million Euros or four percent of an enterprise's prior-year income—enterprises are finding it increasingly important to manage these areas of risk. In the United States, health information is subject to strong protections (the Health Insurance portability and Accountability Act [HIPAA]), while the state of California recently augmented a consumer privacy law similar in intent to GDPR (the California Privacy Rights Act [CPRA]).

The term "data" has traditionally been a subset term used to refer to digital forms of information. In the context of risk management, this distinction between data and information continues to apply; however, risk practitioners should be conscious that laws and regulations that deal with data privacy may also apply to information in non-digital formats. Practitioners should be aware that privacy is a dynamic field subject to regular changes and that rules applicable to certain contexts may conflict with those that apply to other contexts, including geography and types of data. Even in a cloud environment, physical storage of data may impose obligations on an organization, and international agreements are not always reliable over the long term. For instance, the US-EU "Safe Harbor" program established to facilitate compliance with GDPR was declared insufficient by a European court in 2020. Each set of laws and regulations specific to a given context should be given appropriate study for the risk practitioner's role and responsibilities, and care should be taken to maintain currency in those areas as things change.

4.10.1 Key Concepts of Data Privacy

At its essence, privacy is very similar to confidentiality: the goal is to ensure that no one gets access to information who should not have it. What distinguishes privacy from confidentiality in practice is that an individual identified by information subject to privacy controls—the subject—has rights regarding the handling and retention of that data even if that person is not the data owner. This distinction establishes several concepts specific to privacy that deserve the risk practitioner's awareness, which are also reflected in the seven principles of GDPR. An effective privacy program should establish clear lines of responsibility and accountability that include these concepts in a manner appropriate for the legal and business contexts in which the organization operates.

Informed Consent

Data subject to privacy regulations should be collected, used and retained with the informed consent of the subject. A signed consent form is typically part of the process but may not unto itself be enough to establish informed consent. In addition, there may need to be a process for revocation of consent.

Privacy Impact Assessment

Organizations conduct privacy impact assessments (PIAs) to identify and manage risk related to privacy whenever personal information is collected. The PIA typically covers how information is used, shared and maintained. Similar to a risk assessment, a PIA accounts for the effects of a compromise on the subject rather than impact to the enterprise. Risk practitioners dealing with data subject to privacy regulations should be familiar with the PIAs conducted prior to collecting the data as a first step towards ensuring that adequate protective measures are in place.

Minimization

Many enterprises accumulate considerable data and retain it without limit. Risk practitioners should understand that this data, which may be viewed as an asset, is also a potential disaster for enterprises subject to strong privacy regulations should a data breach occur. Privacy is best assured when only data that is expressly needed and relevant for a particular purpose is collected.

Destruction

Just as data should only be collected when necessary, it should generally be destroyed once its purpose is concluded. Destruction can be complicated, because there are some types of data that must be retained for specific periods according to law or regulation, and contractual obligations may also impose specific parameters for destruction. In this case, it is important to implement access restrictions to the data. Risk practitioners should account for destruction requirements and ensure that there are methods to carry out destruction as well as processes to verify that it has occurred as planned.

4.10.2 Risk Management in a Privacy Context

Data privacy is a comprehensive field of study involving multiple principles, frameworks and standards in addition to legal and regulatory requirements that can vary widely from one jurisdiction to another. Penalties for violations of privacy law can be enormous in some jurisdictions. Enterprises in these places will almost certainly have substantial

CHAPTER 4— INFORMATION TECHNOLOGY AND SECURITY

privacy expertise on staff and be prepared to make investments in security commensurate with the potential impact. Risk practitioners assessing systems that handle sensitive personal information should consult with organizational privacy officers or privacy experts to ensure correct measures are in place for the contexts and use cases under consideration.

While legal and regulatory requirements vary, the principles of data protection that apply to confidentiality in other security contexts also tend to apply to privacy. Access to information subject to privacy considerations should be limited to individuals with a valid business need, and the extent of that access should be limited according to the principle of least privilege. Many enterprises segregate duties for release of private data to serve as checks against accidental release. Strong authentication is highly recommended for access to systems hosting data covered by privacy law, and this data should be encrypted at rest and in transit, with positive identification and nonrepudiation of sending authorities.

Where specific jurisdictional security requirements exist, risk practitioners should ensure that these are addressed by appropriate security controls.

APPENDIX A: CRISC General Exam Information

ISACA is a professional membership association composed of individuals interested in IS audit, assurance, control, security, governance and data privacy. The CRISC Certification Working Group is responsible for establishing policies for the CRISC certification program and developing the exam.

Note: Because information regarding the CRISC examination may change, please refer to www.isaca.org/credentialing/crisc for the most current information.

Requirements for Certification

The CRISC designation is awarded to individuals who have met the following requirements: (1) achieve a passing score on the CRISC exam, (2) adhere to the Code of Professional Ethics, (3) adhere to the continuing education policy, and (4) demonstrate the required minimum work experience supporting the risk management function of an enterprise.

Successful Completion of the CRISC Exam

The exam is open to all individuals who wish to take it. Successful exam candidates are not certified until they apply for certification (and demonstrate that they have met all requirements) and receive approval from ISACA.

Experience in Risk

CRISC candidates must meet the stated experience requirements to become certified. Please refer to www.isaca.org/credentialing/crisc for experience requirements and a list of experience waivers.

Experience must have been gained within the 10-year period preceding the application date for certification, or within five years from the date of initially passing the exam. A completed application for certification must be submitted within five years from the passing date of the CRISC exam. All experience must be independently verified with employers.

Description of the Exam

The CRISC Certification Working Group oversees the development of the exam and ensures the currency of its content. Questions for the CRISC exam are developed through a multitiered process designed to enhance the ultimate quality of the exam.

The purpose of the exam is to evaluate a candidate's knowledge and experience in data privacy. The exam consists of 150 multiple-choice questions, administered during a 4-hour session.

Registration for the CRISC Exam

The CDPSE exam is administered on a continuous basis at qualifying test sites Please refer to the ISACA Exam Candidate Information Guide at www.isaca.org/credentialing/crisc for specific exam registration information including registration, scheduling and languages, and important key information for exam day. Exam registrations can be made online at www.isaca.org/credentialing/crisc.

CRISC Program Accreditation Renewed Under ISO/IEC 17024:2012

The American National Standards Institute (ANSI) has voted to continue the accreditation for the CISA, CISM, CGEIT, CRISC and CDPSE certifications, under ISO/IEC 17024:2012, General Requirements for Bodies Operating Certification Systems of Persons. ANSI, a private, nonprofit organization, accredits other organizations to serve as third-party product, system and personnel certifiers.

ISO/IEC 17024 specifies the requirements to be followed by organizations certifying individual against specific requirements. ANSI describes ISO/IEC 17024 as "expected to play a prominent role in facilitating global standardization of the certification community, increasing mobility among countries, enhancing public safety, and protecting consumers."

ANSI's accreditation:

- Promotes the unique qualifications and expertise ISACA's certifications provide
- Protects the integrity of the certifications and provides legal defensibility
- Enhances consumer and public confidence in the certifications and the people who hold them
- Facilitates mobility across borders or industries

Accreditation by ANSI signifies that ISACA's procedures meet ANSI's essential requirements for openness, balance, consensus and due process. With this accreditation, ISACA anticipates that significant opportunities for CISAs, CISMs, CGEITs, CRISCs and CDPSEs will continue to open in the United States and around the world.

Scheduling the Exam

The CRISC exam can be scheduled directly from your My ISACA Certification Dashboard. Please see the Exam Candidate Scheduling Guide for complete instructions. Exams can be scheduled for any available time slot. Exams may be rescheduled a minimum of 48 hours prior to the originally scheduled appointment. If you are within 48 hours of your original appointment, you must take your exam or forfeit the exam registration fee.

Sitting for the Exam

Prior to the day of the exam make sure you:

- Locate the test center and confirm the start time
- Plan to arrive 15 minutes prior to exam start time
- Plan to store personal belongings
- Review the exam day rules

You must present an acceptable form of identification (ID) in order to enter the testing center. Please see the *Exam Candidate Information Guide* for acceptable forms of ID.

You are prohibited from bringing the following into the test center:

- Reference materials, paper, notepads or language dictionaries
- Calculators
- Any type of communication, surveillance or recording devices such as:
  - Mobile phones
  - Tablets

- - Smart watches or glasses
  - Mobile devices
- Baggage of any kind including handbags, purses or briefcases
- Weapons
- Tobacco products
- Food or beverages
- Visitors

If exam candidates are viewed with any such communication, surveillance or recording devices during the exam administration, their exam will be voided, and they will be asked to immediately leave the exam site.

Personal items brought to the testing center must be stored in a locker or other designated area until the exam is completed and submitted.

Avoid activities that would invalidate your test score.

- Creating a disturbance
- Giving or receiving help; using notes, papers or other aids
- Attempting to take the exam for someone else
- Possession of communication, surveillance or recording devices, including but not limited to cell phones, tablets, smart glasses, smart watches, mobile devices, etc., during the exam administration
- Attempting to share test questions or answers or other information contained in the exam (as such are the confidential information of ISACA), including sharing test questions subsequent to the exam
- Leaving the testing area without authorization. (You will not be allowed to return to the testing room.)
- Accessing items stored in the personal belongings area before the completion of the exam

Budgeting Your Time

The exam is administered over a four-hour period. This allows for a little more than 1.5 minutes per question. Therefore, it is advisable that candidates pace themselves to complete the entire exam.

Grading the Exam

Candidate scores are reported as a scaled score. A scaled score is a conversion of a candidate's raw score on an exam to a common scale. ISACA uses and reports scores on a common scale from 200 to 800.

A candidate must receive a score of 450 or higher to pass the exam. A score of 450 represents a minimum consistent standard of knowledge as established by ISACA's CRISC Certification Working Group. A candidate receiving a passing score may then apply for certification if all other requirements are met.

Passing the exam does not grant the CRISC designation. To become a CRISC, each candidate must complete all requirements, including submitting an application and receiving approval for certification.

The CRISC examination contains some questions that are included for research and analysis purposes only. Those questions are not separately identified, and the candidate's final score will be based only on the common scored questions. There are various versions of each exam but only the common questions are scored for your results.

A candidate receiving a score less than 450 is not successful and can retake the exam by registering and paying the appropriate exam fee. To assist with future study, the result letter each candidate receives will include a score analysis by content area.

You will receive a preliminary score on screen immediately following the completion of your exam. **Your official score will be emailed to you and available online within 10 working days**. Question-level results cannot be provided.

In order to become CRISC-certified, candidates must pass the CRISC exam and must complete and submit an application for certification (and must receive confirmation from ISACA that the application is approved). The application is available on the ISACA website at www.isaca.org/credentialing/crisc. Once the application is approved, the applicant will be sent confirmation of the approval. The candidate is not CRISC-certified, and cannot use the CRISC designation, until the candidate's application is approved. A processing fee must accompany CRISC applications for certification.

Candidates receiving a failing score on the exam may request a rescoring of their exam within 30 days following the release of the exam results. All requests must include a candidate's name, exam identification number and mailing address. A fee of US $75 must accompany this request.

APPENDIX B: CRISC Job Practice

Knowledge Subdomains

Domain 1—Governance

A. Organizational Governance

1. Organizational Strategy, Goals and Objectives
2. Organizational Structure, Roles and Responsibilities
3. Organizational Culture
4. Policies and Standards
5. Business Processes
6. Organizational Assets

B. Risk Governance

1. Enterprise Risk Management and Risk Management Frameworks
2. Three Lines of Defense
3. Risk Profile
4. Risk Appetite and Risk Tolerance
5. Legal, Regulatory and Contractual Requirements
6. Professional Ethics of Risk Management

Domain 2—IT Risk Assessment

A. IT Risk Identification

1. Risk Events
2. Threat Modelling and Threat Landscape
3. Vulnerability and Control Deficiency Analysis
4. Risk Scenario Development

B. IT Risk Analysis and Evaluation

1. Risk Assessment Concepts, Standards and Frameworks
2. Risk Register
3. Risk Analysis Methodologies
4. Business Impact Analysis
5. Inherent and Residual Risk

Domain 3—Risk Response and Reporting

A. Risk Response

1. Risk Treatment/Risk Response Options
2. Risk and Control Ownership
3. Third-Party Risk Management
4. Issue, Finding, and Exception Management

5. Management of Emerging Risk

B. Control Design and Implementation

1. Control Types, Standards, and Frameworks
2. Control Design, Selection, and Analysis
3. Control Implementation
4. Control Testing and Effectiveness Evaluation

C. Risk Monitoring and Reporting

1. Risk Treatment Plans
2. Data Collection, Aggregation, Analysis and Validation
3. Risk and Control Monitoring Techniques
4. Risk and Control Reporting Techniques
5. Key Performance Indicators
6. Key Risk Indicators
7. Key Control Indicators

Domain 4—Information Technology and Security

A. Information Technology Principles

1. Enterprise Architecture
2. IT Operations Management
3. Project Management
4. Disaster Recovery Management
5. Data Lifecycle Management
6. System Development Life Cycle
7. Emerging Technologies

B. Information Security Principles

1. Information Security Concepts, Frameworks and Standards
2. Information Security Awareness Training
3. Business Continuity Management
4. Data Privacy and Data Protection Principles

Supporting Task Statements

1. Collect and review existing information regarding the organization's business and IT environments.
2. Identify potential or realized impacts of IT risk to the organization's business objectives and operations.
3. Identify threats and vulnerabilities to the organization's people, processes and technology.
4. Evaluate threats, vulnerabilities, and risk to identify IT risk scenarios.
5. Establish accountability by assigning and validating appropriate levels of risk and control ownership.
6. Establish and maintain the IT risk register, and incorporate it into the enterprise-wide risk profile.
7. Facilitate the identification of risk appetite and risk tolerance by key stakeholders.
8. Promote a risk-aware culture by contributing to the development and implementation of security awareness

training.

9. Conduct a risk assessment by analyzing IT risk scenarios and determining their likelihood and impact.
10. Identify the current state of existing controls and evaluate their effectiveness for IT risk mitigation.
11. Review the results of risk analysis and control analysis to assess any gaps between current and desired states of the IT risk environment.
12. Facilitate the selection of recommended risk responses by key stakeholders.
13. Collaborate with risk owners on the development of risk treatment plans.
14. Collaborate with control owners on the selection, design, implementation and maintenance of controls.
15. Validate that risk responses have been executed according to risk treatment plans.
16. Define and establish key risk indicators (KRIs).
17. Monitor and analyze key risk indicators (KRIs).
18. Collaborate with control owners on the identification of key performance indicators (KPIs) and key control indicators (KCIs).
19. Monitor and analyze key performance indicators (KPIs) and key control indicators (KCIs).
20. Review the results of control assessments to determine the effectiveness and maturity of the control environment.
21. Conduct aggregation, analysis, and validation of risk and control data.
22. Report relevant risk and control information to applicable stakeholders to facilitate risk-based decision-making.
23. Evaluate emerging technologies and changes to the environment for threats, vulnerabilities and opportunities.
24. Evaluate alignment of business practices with risk management and information security frameworks and standards.

Page intentionally left blank

Glossary

Note: Glossary terms are provided for reference within the CRISC Review Manual. As definitions of terms may evolve due to the changing technological environment, please see www.isaca.org/resources/glossary for the most up-to-date terms and definitions.

A

Access control— The processes, rules and deployment mechanisms that control access to information systems, resources and physical access to premises

Access rights— The permission or privileges granted to users, programs or workstations to create, change, delete or view data and files within a system, as defined by rules established by data owners and the information security policy

Access risk— The risk that information may be divulged or made available to recipients without authorized access from the information owner, reflecting a loss of confidentiality

Accountability— The ability to map a given activity or event back to the responsible party

Administrative control— The rules, procedures and practices dealing with operational effectiveness, efficiency and adherence to regulations and management policies.

Advanced persistent threat (APT)— An adversary that possesses sophisticated levels of expertise and significant resources that allow it to create opportunities to achieve its objectives by using multiple attack vectors, including cyber, physical and deception. Typically, APT objectives include establishing and extending footholds within the IT infrastructure of the targeted organizations for purposes of exfiltrating information, or undermining or impeding critical aspects of a mission, program or organization; or positioning itself to carry out those objectives in the future. The advanced persistent threat pursues its objectives repeatedly, over an extended period, adapts to defenders' efforts to resist it and is determined to maintain the level of interaction that is needed to execute its objectives.

Source: NIST SP 800-39

Application controls— The policies, procedures and activities designed to provide reasonable assurance that objectives relevant to a given automated solution (application) are achieved.

Application programming interface (API)— A set of routines, protocols and tools referred to as building blocks used in business application software development

Architecture— Description of the fundamental underlying design of the components of the business system, or of one element of the business system (e.g., technology), the relationships among them, and the manner in which they support enterprise objectives

Artificial intelligence— Advanced computer systems that can simulate human capabilities, such as analysis, based on a predetermined set of rules

Asset— Something of either tangible or intangible value that is worth protecting, including people, information, infrastructure, finances and reputation

Asset inventory— A register that is used to record all relevant assets

Asset value— The value of an asset to both the business and to competitors

Attack— An actual occurrence of an adverse event

Attack mechanism— A method used to deliver the exploit. Unless the attacker is personally performing the attack, an attack mechanism may involve a payload, or container, that delivers the exploit to the target.

Attack vector— A path or route used by the adversary to gain access to the target (asset)

Scope Notes: There are two types of attack vectors: ingress and egress (also known as data exfiltration).

Authentication— 1. The act of verifying identity, i.e., user, system

Scope Notes: Can also refer to the verification of the correctness of a piece of data.

2. The act of verifying the identity of a user, the user's eligibility to access computerized information

Scope Notes: Authentication is designed to protect against fraudulent logon activity. It can also refer to the verification of the correctness of a piece of data.

Authenticity— Undisputed authorship

Authorization— The process of determining if the end user is permitted to have access to an information asset or the information system containing the asset.

GLOSSARY

Availability— Ensuring timely and reliable access to and use of information

Availability risk— The risk that service may be lost or data are not accessible when needed

Awareness— Being acquainted with, mindful of, conscious of and well informed on a specific subject, which implies knowing and understanding a subject and acting accordingly.

B

Balanced scorecard (BSC)— Developed by Robert S. Kaplan and David P. Norton as a coherent set of performance measures organized into four categories that includes traditional financial measures, but adds customer, internal business process, and learning and growth perspectives.

Big data— The ability to work with collections of data that had been impractical previously because of their volume, velocity and variety (the three Vs). A key driver of big-data ability was easier distribution of storage and processing across networks of inexpensive commodity hardware, using technology such as Hadoop, instead of requiring larger, more powerful individual computers.

Biometrics— A security technique that verifies an individual's identity by analyzing a unique physical attribute, such as a handprint

Bring your own device (BYOD)— An enterprise policy used to permit partial or full integration of user-owned mobile devices for business purposes

Business case— Documentation of the rationale for making a business investment, used both to support a business decision on whether to proceed with the investment and as an operational tool to support management of the investment through its full economic life cycle

Business continuity— Preventing, mitigating and recovering from disruption

Scope Notes: The terms 'business resumption planning', 'disaster recovery planning' and 'contingency planning' also may be used in this context; they focus on recovery aspects of continuity, and for that reason the 'resilience' aspect should also be taken into account.

COBIT 5 and COBIT 2019 perspective

Business continuity plan (BCP)— A plan used by an enterprise to respond to disruption of critical business processes; depends on the contingency plan for restoration of critical systems

Business control— The policies, procedures, practices and organizational structures designed to provide reasonable assurance that the business objectives will be achieved and undesired events will be prevented or detected.

Business goal— The translation of the enterprise's mission from a statement of intention into performance targets and results.

Business impact— The net effect, positive or negative, on the achievement of business objectives.

Business impact analysis (BIA)— Process of evaluating the criticality and sensitivity of information assets by determining the impact of losing the support of any resource to an enterprise. Establishes the escalation of that loss over time, identifies the minimum resources needed to recover and prioritizes the recovery of processes and the supporting system.

Scope Notes: This process captures income loss, unexpected expense, legal issues (regulatory compliance or contractual), interdependent processes and loss of public reputation or public confidence.

Business objective— A further development of the business goals into tactical targets and desired results and outcomes.

Business process— An inter-related set of cross-functional activities or events that result in the delivery of a specific product or service to a customer.

Business process owner— The individual responsible for identifying process requirements, approving process design and managing process performance.

Scope Notes: Must be at an appropriately high level in the enterprise and have authority to commit resources to process-specific risk management activities

Business risk— The probability that a situation with uncertain frequency and magnitude of loss (or gain) could prevent the enterprise from meeting its business objectives

C

Capability— An aptitude, competency or resource that an enterprise may possess or require at an enterprise, business function or individual level that has the potential, or is required, to contribute to a business outcome and to create value.

Capability Maturity Model (CMM)— 1. Contains the essential elements of effective processes for one or more disciplines. It also describes an evolutionary improvement path from ad hoc, immature processes to disciplined, mature processes with improved quality and effectiveness.

2. CMM for software, from the Software Engineering Institute (SEI), is a model used by many enterprises to identify best practices useful in helping them assess and increase the maturity of their software development processes.

Scope Notes: CMM ranks software development enterprises according to a hierarchy of five process maturity levels. Each level ranks the development environment according to its capability of producing quality software. A set of standards is associated with each of the five levels. The standards for level one describe the most immature or chaotic processes and the standards for level five describe the most mature or quality processes. A maturity model that indicates the degree of reliability or dependency the business can place on a process achieving the desired goals or objectives. A collection of instructions that an enterprise can follow to gain better control over its software development process.

Certificate (Certification) authority (CA)— A trusted third party that serves authentication infrastructures or enterprises and registers entities and issues them certificates

Certificate revocation list (CRL)— An instrument for checking the continued validity of the certificates for which the certification authority (CA) has responsibility

Scope Notes: The CRL details digital certificates that are no longer valid. The time gap between two updates is very critical and is also a risk in digital certificates verification.

Change control— The processes, authorities and procedures to be used for all changes that are made to the computerized system and/or the system data. Change control is a vital subset of the quality assurance (QA) program in an enterprise and should be clearly described in the enterprise standard operating procedures (SOPs).

See Configuration control.

Change management— A holistic and proactive approach to managing the transition from a current to a desired organizational state, focusing specifically on the critical human or "soft" elements of change.

Scope Notes: Includes activities such as culture change (values, beliefs and attitudes), development of reward systems (measures and appropriate incentives), organizational design, stakeholder management, human resources (HR) policies and procedures, executive coaching, change leadership training, team building and communication planning and execution.

Change risk— A change in technology, regulation, business process, functionality, architecture, user and other variables that affect the enterprise business and technical environments, and the level of risk associated with systems in operation

Cloud computing— Convenient, on-demand network access to a shared pool of resources that can be rapidly provisioned and released with minimal management effort or service provider interaction

Code of ethics— A document designed to influence individual and organizational behavior of employees, by defining organizational values and the rules to be applied in certain situations.

Scope Notes: A code of ethics is adopted to assist those in the enterprise called upon to make decisions understand the difference between 'right' and 'wrong' and to apply this understanding to their decisions.

COBIT 5 and COBIT 2019 perspective

Compensating control— An internal control that reduces the risk of an existing or potential control weakness resulting in errors and omissions.

Compliance risk— The probability and consequences of an enterprise failing to comply with laws, regulations or the ethical standards or codes of conduct applicable to the enterprise industry

Computer emergency response team (CERT)— A group of people integrated at the enterprise with clear lines of reporting and responsibilities for standby support in case of an information systems emergency. This group will act as an efficient corrective control, and should also act as a single point of contact for all incidents and issues related to information systems.

Confidentiality— Preserving authorized restrictions on access and disclosure, including means for protecting privacy and proprietary information

GLOSSARY

Configuration management— The control of changes to a set of configuration items over a system life cycle

Consequence— The result of a realized risk. A consequence can be certain or uncertain and can have positive or negative direct or indirect effects on objectives. Consequences can be expressed qualitatively or quantitatively.

Continuous risk and control monitoring— A process that includes:

- Developing a strategy to regularly evaluate selected information and technology (I&T)-related controls/metrics
- Recording and evaluating I&T-related events and the effectiveness of the enterprise in dealing with those events
- Recording changes to I&T-related controls or changes that affect I&T-related risk
- Communicating the current risk and control status to enable information-sharing decisions involving the enterprise

Control— The means of managing risk, including policies, procedures, guidelines, practices or organizational structures, which can be of an administrative, technical, management or legal nature

Scope Notes: Also used as a synonym for safeguard or countermeasure.

See also Internal control.

Control objective— A statement of the desired result or purpose to be achieved by implementing control procedures in a particular process.

Control owner— A person in whom the enterprise has invested the authority and accountability for making control-related decisions and is responsible for ensuring that the control is implemented and is operating effectively and efficiently

Control risk— Risk that assets are lost/compromised or that financial statements are materially misstated due to lack of or ineffective design and/or implementation of internal controls

Control risk self-assessment— A method/process by which management and staff of all levels collectively identify and evaluate risk and controls with their business areas. This may be under the guidance of a facilitator such as an auditor or risk manager.

Control self-assessment— An assessment of controls made by the staff and management of the unit or units involved

Control weakness— A deficiency in the design or operation of a control procedure. Control weaknesses can potentially result in risk relevant to the area of activity not being reduced to an acceptable level (relevant risk threatens achievement of the objectives relevant to the area of activity being examined). Control weaknesses can be material when the design or operation of one or more control procedures does not reduce to a relatively low level the risk that misstatements caused by illegal acts or irregularities may occur and not be detected by the related control procedures.

Copyright— Protection of writings, recordings or other ways of expressing an idea. The idea itself may be common, but they way it was expressed is unique, such as a song or book

Culture— A pattern of behaviors, beliefs, assumptions, attitudes and ways of doing things

Scope Notes: COBIT 5 and COBIT 2019 perspective

Current risk— The risk state that exists in the moment, taking into account those actions that have already been taken but not actions that are anticipated or have been proposed

Cyber and information security risk— The danger, harm or loss related to the use of, or dependence on, information and communications technology, electronic data, and digital or electronic communications

D

Data classification— The assignment of a level of sensitivity to data (or information) that results in the specification of controls for each level of classification. Levels of sensitivity of data are assigned according to predefined categories as data are created, amended, enhanced, stored or transmitted. The classification level is an indication of the value or importance of the data to the enterprise.

Data classification scheme— An enterprise scheme for classifying data by factors such as criticality, sensitivity and ownership.

Data custodian— The individual(s) and department(s) responsible for the storage and safeguarding of computerized data.

GLOSSARY

Data destruction— The elimination, erasure or clearing of data

Data owner— The individual(s) who has responsibility for the integrity, accurate reporting and use of computerized data

Decryption— A technique used to recover the original plaintext from the ciphertext so that it is intelligible to the reader. The decryption is a reverse process of the encryption.

Demilitarized zone (DMZ)— A small, isolated network that serves as a buffer zone between trusted and untrusted networks

Scope Notes: A DMZ is typically used to house systems, such as web servers, that must be accessible from both internal networks and the Internet.

Detection risk— Risk that assets are lost/compromised or that financial statements are materially misstated due to failure of an enterprise's internal controls to detect errors or fraud in a timely manner

Detective control— Designed to detect and report when errors, omissions and unauthorized uses or entries occur

Digital certificate— Electronic credentials that permit an entity to exchange information securely via the Internet using the public key infrastructure (PKI)

Digital signature— An electronic identification of a person or entity using a public key algorithm that serves as a way for the recipient to verify the identity of the sender, integrity of the data and proof of transaction

Disaster recovery— Activities and programs designed to return the enterprise to an acceptable condition. The ability to respond to an interruption in services by implementing a disaster recovery plan (DRP) to restore an enterprise's critical business functions.

Disaster recovery plan (DRP)— A set of human, physical, technical and procedural resources to recover, within a defined time and cost, an activity interrupted by an emergency or disaster

E

Encryption— The process of taking an unencrypted message (plaintext), applying a mathematical function to it (encryption algorithm with a key) and producing an encrypted message (ciphertext).

Encryption algorithm— A mathematically based function or calculation that encrypts/decrypts data; may be block or stream ciphers.

Encryption key— A piece of information, in a digitized form, used by an encryption algorithm to convert the plaintext to the ciphertext

Enterprise risk management (ERM)— The discipline by which an enterprise in any industry assesses, controls, exploits, finances and monitors risk from all sources for the purpose of increasing the enterprise's short- and long-term value to its stakeholders.

Environmental risk— Threats to natural resources, human health and wildlife

Event— Something that happens at a specific place and/or time

Event type— For the purpose of IT risk management, one of three possible sorts of events: threat event, loss event and vulnerability event.

Scope Notes: Being able to consistently and effectively differentiate the different types of events that contribute to risk is a critical element in developing good risk-related metrics and well-informed decisions. Unless these categorical differences are recognized and applied, any resulting metrics lose meaning and, as a result, decisions based on those metrics are far more likely to be flawed.

Evidence— 1. Information that proves or disproves a stated issue

2. Information that an auditor gathers in the course of performing an IS audit; relevant if it pertains to the audit objectives and has a logical relationship to the findings and conclusions it is used to support

Scope Notes: Audit perspective

Exploit— Method used to take advantage of a vulnerability

F

Fallback procedures— A plan of action or set of procedures to be performed if a system implementation, upgrade or modification does not work as intended.

Scope Notes: May involve restoring the system to its state prior to the implementation or change. Fallback procedures are needed to ensure that normal business processes continue in the event of failure and should

GLOSSARY

always be considered in system migration or implementation.

Feasibility study— Analysis of the known or anticipated need for a product, system or component to assess the degree to which the requirements, designs or plans can be implemented

Firewall— A system or combination of systems that enforces a boundary between two or more networks, typically forming a barrier between a secure and an open environment such as the Internet

Framework— A framework is a basic conceptual structure used to solve or address complex issues. An enabler of governance. A set of concepts, assumptions and practices that define how something can be approached or understood, the relationships among the entities involved, the roles of those involved and the boundaries (what is and is not included in the governance system).

See Control framework and IT governance framework.

Frequency— A measure of the rate by which events occur over a certain period of time

G

Governance— The method by which an enterprise ensures that stakeholder needs, conditions and options are evaluated to determine balanced, agreed-on enterprise objectives are achieved. It involves setting direction through prioritization and decision making, and monitoring performance and compliance against agreed-on direction and objectives.

Governance component— Factors that, individually and collectively, contribute to the good operation of the enterprise's governance system over information and technology (I&T). Components interact with each other resulting in a holistic governance system for I&T. Components include processes; organizational structures; principles, policies and procedures; information; culture, ethics and behavior; people, skills and competencies; and services, infrastructure and applications.

Governance of enterprise IT— A governance view that ensures that information and related technology support and enable the enterprise strategy and the achievement of enterprise objectives; this also includes the functional governance of IT, i.e., ensuring that IT capabilities are provided efficiently and effectively.

Scope Notes: COBT 5 perspective

Governance, risk management and compliance (GRC)— A business term used to group the three closely related disciplines responsible for operations and the protection of assets.

I

Impact— Magnitude of loss resulting from a threat exploiting a vulnerability

Impact analysis— A study to prioritize the criticality of information resources for the enterprise based on costs (or consequences) of adverse events. In an impact analysis, threats to assets are identified and potential business losses determined for different time periods. This assessment is used to justify the extent of safeguards that are required and recovery time frames. This analysis is the basis for establishing the recovery strategy.

Impact assessment— A review of the possible consequences of a risk.

Scope Notes: See also Impact analysis.

Incident— A violation or imminent threat of violation of computer security policies, acceptable use policies, guidelines or standard security practices

Incident response— The response of an enterprise to a disaster or other significant event that may significantly affect the enterprise, its people or its ability to function productively. An incident response may include evacuation of a facility, initiating a disaster recovery plan (DRP), performing damage assessment and any other measures necessary to bring an enterprise to a more stable status.

Incident response plan— Also called IRP. The operational component of incident management.

Scope Notes: The plan includes documented procedures and guidelines for defining the criticality of incidents, reporting and escalation process, and recovery procedures.

Information and technology (I&T)-related risk— A part of overall business risk associated with the use, ownership, operation, involvement, influence and adoption of information and technology (I&T) within an enterprise

Information security— Ensures that, within the enterprise, information is protected against disclosure to unauthorized users (confidentiality), improper modification (integrity) and nonaccess when required (availability). Information security deals with all formats of information – paper documents, digital assets, intellectual property in people's minds, and verbal and visual communications.

Information systems (IS)— The combination of strategic, managerial and operational activities involved in gathering, processing, storing, distributing and using information and its related technologies

Scope Notes: Information systems are distinct from information technology (IT) in that an information system has an IT component that interacts with the process components.

Information technology (IT)— The hardware, software, communication and other facilities used to input, store, process, transmit and output data in whatever form.

Infrastructure as a Service (IaaS)— Offers the capability to provision processing, storage, networks and other fundamental computing resources, enabling the customer to deploy and run arbitrary software, which can include operating systems (OSs) and applications

Inherent risk— The risk level or exposure without taking into account the actions that management has taken or might take (e.g., implementing controls)

Integrity— The guarding against improper information modification or destruction, and includes ensuring information nonrepudiation and authenticity

Intellectual property— Intangible assets that belong to an enterprise for its exclusive use. Examples include patents, copyrights, trademarks, ideas and trade secrets.

Internal control environment— The relevant environment on which the controls have effect.

Internal controls— The policies, procedures, practices and organizational structures designed to provide reasonable assurance that business objectives will be achieved and undesired events will be prevented or detected and corrected.

IT architecture— Description of the fundamental underlying design of the IT components of the business, the relationships among them, and the manner in which they support the enterprise's objectives.

IT infrastructure— The set of hardware, software and facilities that integrates an enterprise's IT assets.

Scope Notes: Specifically, the equipment (including servers, routers, switches and cabling), software, services and products used in storing, processing, transmitting and displaying all forms of information for the enterprise's users

IT risk— The business risk associated with the use, ownership, operation, involvement, influence and adoption of IT within an enterprise.

IT risk issue— 1. An instance of IT risk.

2. A combination of control, value and threat conditions that impose a noteworthy level of IT risk.

IT risk profile— A description of the overall (identified) IT risk to which the enterprise is exposed.

IT risk register— A repository of the key attributes of potential and known IT risk issues. Attributes may include name, description, owner, expected/actual frequency, potential/actual magnitude, potential/actual business impact, disposition.

IT risk scenario— The description of an IT-related event that can lead to a business impact.

IT strategic plan— A long-term plan (i.e., three- to five-year horizon) in which business and IT management cooperatively describe how IT resources will contribute to the enterprise's strategic objectives (goals).

IT tactical plan— A medium-term plan (i.e., six- to 18-month horizon) that translates the IT strategic plan direction into required initiatives, resource requirements and ways in which resources and benefits will be monitored and managed.

IT-related incident— An IT-related event that causes an operational, developmental and/or strategic business impact.

K

Key control indicator (KCI)— A measure of the effectiveness of controls to indicate the failure or weakness which may result in the increase of the likelihood or impact of risk events.

Key performance indicator (KPI)— A measure that determines how well the process is performing in enabling the goal to be reached

Scope Notes: A lead indicator of whether a goal will likely be reached, and a good indicator of capabilities, practices and skills. It measures an activity goal, which is an action that the process owner must take to achieve effective process performance.

GLOSSARY

Key risk indicator (KRI)— A subset of risk indicators that are highly relevant and possess a high probability of predicting or indicating important risk

Scope Notes: See also Risk indicator.

L

Lag risk indicator— A backward-looking metric that indicates risk has been realized after an event has occurred

Lead risk indicator— A lead risk indicator is a forward-looking metric that provides an early warning that risk may soon be realized before an event has occurred.

Likelihood— The probability of something happening

Loss event— Any event during which a threat event results in loss.

Scope Notes: From Jones, J.; "FAIR Taxonomy," Risk Management Insight, USA, 2008

M

Magnitude— A measure of the potential severity of loss or the potential gain from realized events/scenarios

Management— Plans, builds, runs and monitors activities in alignment with the direction set by the governance body to achieve the enterprise objectives.

Market risk— Pressures on an asset class

N

Nondisclosure agreement (NDA)— A legal contract between at least two parties that outlines confidential materials that the parties wish to share with one another for certain purposes, but wish to restrict from generalized use; a contract through which the parties agree not to disclose information covered by the agreement.

Scope Notes: Also called a confidential disclosure agreement (CDA), confidentiality agreement or secrecy agreement. An NDA creates a confidential relationship between the parties to protect any type of trade secret. As such, an NDA can protect non-public business information. In the case of certain governmental entities, the confidentiality of information other than trade secrets may be subject to applicable statutory requirements, and in some cases may be required to be revealed to an outside party requesting the information. Generally, the governmental entity will include a provision in the contract to allow the seller to review a request for information that the seller identifies as confidential and the seller may appeal such a decision requiring disclosure. NDAs are commonly signed when two companies or individuals are considering doing business together and need to understand the processes used in one another's businesses solely for the purpose of evaluating the potential business relationship. NDAs can be "mutual," meaning that both parties are restricted in their use of the materials provided, or they can only restrict a single party. It is also possible for an employee to sign an NDA or NDA-like agreement with a company at the time of hiring; in fact, some employment agreements will include a clause restricting "confidential information" in general.

Nonrepudiation— The assurance that a party cannot later deny originating data; provision of proof of the integrity and origin of the data and that can be verified by a third party

Scope Notes: A digital signature can provide nonrepudiation.

O

Objectivity— The ability to exercise judgment, express opinions and present recommendations with impartiality

Operational level agreement (OLA)— An internal agreement covering the delivery of services that support the IT organization in its delivery of services.

Operational risk— The potential for losses caused by inadequate systems or controls, human error or mismanagement, and natural disasters

Owner— Individual or group that holds or possesses the rights of and the responsibilities for an enterprise, entity or asset.

Scope Notes: Examples: process owner, system owner

COBIT 5 perspective

P

Patent— Protection of research and ideas that led to the development of a new, unique and useful product to prevent the unauthorized duplication of the patented item

Penetration testing— A live test of the effectiveness of security defenses through mimicking the actions of real-life attackers

Performance indicators— A set of metrics designed to measure the extent to which performance objectives are being achieved on an on-going basis.

Scope Notes: Performance indicators can include service level agreements (SLAs), critical success factors (CSFs), customer satisfaction ratings, internal or external benchmarks, industry best practices and international standards.

Personally identifiable information (PII)— Any information that can be used to establish a link between the information and the natural person to whom such information relates, or that is or might be directly or indirectly linked to a natural person

Platform as a Service (PaaS)— Offers the capability to deploy onto the cloud infrastructure customer-created or -acquired applications that are created using programming languages and tools supported by the provider

Policy— A document that communicates required and prohibited activities and behaviors

Portfolio— A grouping of "objects of interest" (investment programs, IT services, IT projects, other IT assets or resources) managed and monitored to optimize business value. (The investment portfolio is of primary interest to Val IT. IT service, project, asset and other resource portfolios are of primary interest to COBIT.).

Preventive control— An internal control that is used to avoid undesirable events, errors and other occurrences that an enterprise has determined could have a negative material effect on a process or end product.

Privacy by design— The integration of privacy into the entire engineering process

Privacy controls— Measures that treat privacy risk by reducing its likelihood or consequences. Privacy controls include organizational, physical and technical measures, e.g., policies, procedures, guidelines, legal contracts, management practices or organizational structures. Control is also used as a synonym for safeguard or countermeasure.

Privacy risk— Any risk of informational harm to data subjects and/or organization(s), including deception, financial injury, health and safety injuries, unwanted intrusion, and reputational injuries which harm (or damage) that goes beyond economic and tangible losses

Privilege— The level of trust with which a system object is imbued.

Probability— A mathematical-driven measure of the possibility of a specific outcome as a ratio of all possible outcomes

Problem— In IT, the unknown underlying cause of one or more incidents.

Problem escalation procedure— The process of escalating a problem up from junior to senior support staff, and ultimately to higher levels of management.

Scope Notes: Problem escalation procedure is often used in help desk management, when an unresolved problem is escalated up the chain of command, until it is solved.

Program (Project Management)— A structured grouping of interdependent projects that is both necessary and sufficient to achieve a desired business outcome and create value. These projects could include, but are not limited to, changes in the nature of the business, business processes and the work performed by people as well as the competencies required to carry out the work, the enabling technology, and the organizational structure.

Project— A structured set of activities concerned with delivering a defined capability (that is necessary but not sufficient, to achieve a required business outcome) to the enterprise based on an agreed-on schedule and budget.

Project portfolio— The set of projects owned by a company.

Scope Notes: It usually includes the main guidelines relative to each project, including objectives, costs, time lines and other information specific to the project.

Project risk— A failed IT project that poses a significant risk to an enterprise, manifesting as lost market share, failure to seize new opportunities or other adverse impacts on customers, shareholders and staff

Public key infrastructure (PKI)— A series of processes and technologies for the association of cryptographic keys with the entity to whom those keys were issued

Q

Qualitative risk analysis— Approach based on expert opinion, judgement, intuition and experience

Quantitative risk analysis— Approach that is based on a calculation of a risk's likelihood and impact using numerical and statistical techniques

R

RACI chart— Illustrates who is Responsible, Accountable, Consulted and Informed within an organizational framework.

GLOSSARY

Recovery point objective (RPO)— Determined based on the acceptable data loss in case of a disruption of operations. It indicates the earliest point in time that is acceptable to recover the data. The RPO effectively quantifies the permissible amount of data loss in case of interruption.

Recovery strategy— An approach by an enterprise that will ensure its recovery and continuity in the face of a disaster or other major outage.

Scope Notes: Plans and methodologies are determined by the enterprise's strategy. There may be more than one methodology or solution for an enterprise's strategy. Examples of methodologies and solutions include: contracting for hot site or cold site, building an internal hot site or cold site, identifying an alternate work area, a consortium or reciprocal agreement, contracting for mobile recovery or crate and ship, and many others.

Recovery testing— A test to check the system's ability to recover after a software or hardware failure.

Recovery time objective (RTO)— The amount of time allowed for the recovery of a business function or resource after a disaster occurs

Reputation risk— The current and prospective effect on earnings and capital arising from negative public opinion.

Scope Notes: Reputation risk affects a bank's ability to establish new relationships or services, or to continue servicing existing relationships. It may expose the bank to litigation, financial loss or a decline in its customer base. A bank's reputation can be damaged by Internet banking services that are executed poorly or otherwise alienate customers and the public. An Internet bank has a greater reputation risk as compared to a traditional brick-and-mortar bank, because it is easier for its customers to leave and go to a different Internet bank and since it cannot discuss any problems in person with the customer.

Residual risk— The remaining risk after management has implemented a risk response

Resilience— The ability of a system or network to resist failure or to recover quickly from any disruption, usually with minimal recognizable effect

Return on investment (ROI)— A measure of operating performance and efficiency, computed in its simplest form by dividing net income by the total investment over the period being considered

Risk— The combination of the likelihood of an event and its impact

Risk acceptance— Decision to accept a risk, made according to the risk appetite and risk tolerance set by senior management where the enterprise can assume the risk and absorb any losses

Risk aggregation— The process of integrating risk assessments at a corporate level to obtain a complete view on the overall risk for the enterprise.

Risk analysis— 1. A process by which frequency and magnitude of IT risk scenarios are estimated.

2. The initial steps of risk management: analyzing the value of assets to the business, identifying threats to those assets and evaluating how vulnerable each asset is to those threats.

Scope Notes: It often involves an evaluation of the probable frequency of a particular event, as well as the probable impact of that event.

Risk appetite— The amount of risk, on a broad level, that an entity is willing to accept in pursuit of its mission.

Risk assessment— A process used to identify and evaluate risk and its potential effects

Scope Notes: Risk assessments are used to identify those items or areas that present the highest risk, vulnerability or exposure to the enterprise for inclusion in the IS annual audit plan.

Risk assessments are also used to manage the project delivery and project benefit risk.

Risk avoidance— The process for systematically avoiding risk, constituting one approach to managing risk

Risk awareness program— A program that creates an understanding of risk, risk factors and the various types of risk that an enterprise faces

Risk capacity— The objective magnitude or amount of loss that an enterprise can tolerate without risking its continued existence

Risk culture— The set of shared values and beliefs that governs attitudes toward risk-taking, care and integrity, and determines how openly risk and losses are reported and discussed.

GLOSSARY

Risk evaluation— The process of comparing the estimated risk against given risk criteria to determine the significance of the risk. [ISO/IEC Guide 73:2002].

Risk factor— A condition that can influence the frequency and/or magnitude and, ultimately, the business impact of IT-related events/scenarios

Risk gap— A gap that exists when the acceptable level of risk and the current state of risk are different

Risk identification— The process for determining and documenting the risk an enterprise faces

Risk indicator— A metric capable of showing that the enterprise is subject to, or has a high probability of being subject to, a risk that exceeds the defined risk appetite

Risk management— 1. The coordinated activities to direct and control an enterprise with regard to risk

Scope Notes: In the International Standard, the term "control" is used as a synonym for "measure." (ISO/IEC Guide 73:2002)

2. One of the governance objectives. Entails recognizing risk; assessing the impact and likelihood of that risk; and developing strategies, such as avoiding the risk, reducing the negative effect of the risk and/or transferring the risk, to manage it within the context of the enterprise's risk appetite.

Scope Notes: COBIT 5 perspective

Risk map— A (graphic) tool for ranking and displaying risk by defined ranges for frequency and magnitude.

Risk owner— The person in whom the organization has invested the authority and accountability for making risk-based decisions and who owns the loss associated with a realized risk scenario.

Scope Notes: The risk owner may not be responsible for the implementation of risk treatment.

Risk portfolio view— 1. A method to identify interdependencies and interconnections among risk, as well as the effect of risk responses on multiple types of risk.

2. A method to estimate the aggregate impact of multiple types of risk (e.g., cascading and coincidental threat types/scenarios, risk concentration/correlation across silos) and the potential effect of risk response across multiple types of risk.

Risk reduction— The implementation of controls or countermeasures to reduce the likelihood or impact of a risk to a level within the organization's risk tolerance

Risk register— A list of risk that have been identified, analyzed and prioritized

Risk response— Risk avoidance, risk acceptance, risk sharing/transfer, risk mitigation, leading to a situation that as much future residual risk (current risk with the risk response defined and implemented) as possible (usually depending on budgets available) falls within risk appetite limits.

Risk scenario— The tangible and assessable representation of risk.

Scope Notes: One of the key information items needed to identify, analyze and respond to risk (COBIT 2019 objective APO12)

Risk sharing— **Scope Notes:** See risk transfer

Risk statement— A description of the current conditions that may lead to the loss; and a description of the loss Source: Software Engineering Institute (SEI)

Scope Notes: For a risk to be understandable, it must be expressed clearly. Such a treatment must include a description of the current conditions that may lead to the loss; and a description of the loss.

Risk taxonomy— A scheme for classifying sources and categories of risk that provides a common language for discussing and communicating risk to stakeholders

Risk tolerance— The acceptable level of variation that management is willing to allow for any particular risk as the enterprise pursues its objectives

Risk transfer— The process of assigning risk to another enterprise, usually through the purchase of an insurance policy or by outsourcing the service

Scope Notes: Also known as risk sharing

Risk treatment— The process of selection and implementation of measures to modify risk (ISO/IEC Guide 73:2002)

Risk universe— Encompasses the overall risk environment, defines the areas that risk management activities will address and provides a structure for information and technology (I&T)-related risk management

Root cause analysis— A process of diagnosis to establish the origins of events, which can be used for learning from consequences, typically from errors and problems

S

Scope creep— Also called requirement creep, this refers to uncontrolled changes in a project's scope.

Scope Notes: Scope creep can occur when the scope of a project is not properly defined, documented and controlled. Typically, the scope increase consists of either new products or new features of already approved products. Hence, the project team drifts away from its original purpose. Because of one's tendency to focus on only one dimension of a project, scope creep can also result in a project team overrunning its original budget and schedule. For example, scope creep can be a result of poor change control, lack of proper identification of what products and features are required to bring about the achievement of project objectives in the first place, or a weak project manager or executive sponsor.

Segregation/separation of duties (SoD)— A basic internal control that prevents or detects errors and irregularities by assigning to separate individuals the responsibility for initiating and recording transactions and for the custody of assets.

Scope Notes: Segregation/separation of duties is commonly used in large IT organizations so that no single person is in a position to introduce fraudulent or malicious code without detection.

Service level agreement (SLA)— An agreement, preferably documented, between a service provider and the customer(s)/user(s) that defines minimum performance targets for a service and how they will be measured

Slack time (float)— Time in the project schedule, the use of which does not affect the project's critical path; the minimum time to complete the project based on the estimated time for each project segment and their relationships.

Scope Notes: Slack time is commonly referred to as "float" and generally is not "owned" by either party to the transaction.

Social engineering— An attack based on deceiving users or administrators at the target site into revealing confidential or sensitive information

Software as a service (SaaS)— Offers the capability to use the provider's applications running on cloud infrastructure. The applications are accessible from various client devices through a thin client interface, such as a web browser (e.g., web-based email).

Standard— A mandatory requirement, code of practice or specification approved by a recognized external standards organization, such as International Organization for Standardization (ISO).

Strategic planning— The process of deciding on the enterprise's objectives, on changes in these objectives, and the policies to govern their acquisition and use.

Strategic risk— The risk associated with the future business plans and strategies of an enterprise

System development life cycle (SDLC)— The phases deployed in the development or acquisition of a software system.

Scope Notes: SDLC is an approach used to plan, design, develop, test and implement an application system or a major modification to an application system. Typical phases of SDLC include the feasibility study, requirements study, requirements definition, detailed design, programming, testing, installation and post-implementation review, but not the service delivery or benefits realization activities.

T

Threat— Anything (e.g., object, substance, human) that is capable of acting against an asset in a manner that can result in harm

Scope Notes: A potential cause of an unwanted incident (ISO/IEC 13335)

Threat agent— Methods and things used to exploit a vulnerability

Scope Notes: Examples include determination, capability, motive and resources.

Threat analysis— An evaluation of the type, scope and nature of events or actions that can result in adverse consequences; identification of the threats that exist against enterprise assets

Scope Notes: The threat analysis usually defines the level of threat and the likelihood of it materializing.

Threat event— Any event during which a threat element/actor acts against an asset in a manner that has the potential to directly result in harm

Threat vector— The path or route used by the adversary to gain access to the target

Trademark— A sound, color, logo, saying or other distinctive symbol that is closely associated with a certain product or company

V

Vulnerability— A weakness in the design, implementation, operation or internal control of a process that could expose the system to adverse threats from threat events

Vulnerability analysis— A process of identifying and classifying vulnerabilities

Vulnerability event— Any event during which a material increase in vulnerability results. Note that this increase in vulnerability can result from changes in control conditions or from changes in threat capability/force.

Scope Notes: From Jones, J.; "FAIR Taxonomy," Risk Management Insight, USA, 2008

Vulnerability scanning— An automated process to proactively identify security weaknesses in a network or individual system